Brain–Computer Interfaces

Neurorehabilitation of Voluntary Movement after Stroke and Spinal Cord Injury

Synthesis Lectures on Assistive, Rehabilitative, and Health-Preserving Technologies

Editors
Ronald M. Baecker, *University of Toronto*
Andrew Sixsmith, *Simon Fraser University and AGE-WELL NCE*

This series provides state-of-the-art overview lectures on assistive technologies. We take a broad view of this expanding field, defining it as information and communications technologies used in diagnosis and treatment, prosthetics that compensate for impaired capabilities, methods for rehabilitating or restoring function, and protective interventions that enable individuals to stay healthy for longer periods of time.

Each overview introduces the medical context in which technology is used, presents and explains the technology; reviews problems and opportunities, successes and failures in the development and use of technology; and synthesizes promising opportunities for future progress. Authors include significant material based on their own work, while surveying the broad landscape of an area's research, development, and deployment progress and success.

Brain–Computer Interfaces for Neurorehabilitation of Voluntary Movement after Stroke and Spinal Cord Injury
Cesar Marquez-Chin, Naaz Kapadia-Desai, and Sukhvinder Kalsi-Ryan

ISBN: 978-3-031-00480-3   print
ISBN: 978-3-031-01608-0   ebook
ISBN: 978-3-031-00042-3   hardcover

DOI 10.1007/978-3-031-01608-0

SYNTHESIS LECTURES ON ASSISTIVE, REHABILITATIVE, AND HEALTH-PRESERVING TECHNOLOGIES
Lecture #17

Series Editors: Ron Baecker, University of Toronto and Andrew Sixsmith, Simon Fraser University and AGE-WELL NCE

Series ISSN 2162-7258 Print 2162-7266 Electronic

Brain–Computer Interfaces

Neurorehabilitation of Voluntary Movement after Stroke and Spinal Cord Injury

Cesar Marquez-Chin
KITE Research Institute, Toronto Rehabilitation Institute—University Health Network and University of Toronto

Naaz Kapadia-Desai
KITE Research Institute, Toronto Rehabilitation Institute—University Health Network and University of Toronto

Sukhvinder Kalsi-Ryan
KITE Research Institute, Toronto Rehabilitation Institute—University Health Network and University of Toronto

SYNTHESIS LECTURES ON ASSISTIVE, REHABILITATIVE, AND HEALTH-PRESERVING TECHNOLOGIES #17

ABSTRACT

Stroke and spinal cord injury often result in paralysis with serious negative consequences to the independence and quality of life of those who sustain them. For these individuals, rehabilitation provides the means to regain lost function. Rehabilitation following neurological injuries has undergone revolutionary changes, enriched by neuroplasticity. Neuroplastic-based interventions enhance the efficacy and continue to guide the development of new rehabilitation strategies. This book presents three important technology-based rehabilitation interventions that follow the concepts of neuroplasticity. The book also discusses clinical results related to their efficacy. These interventions are: functional electrical stimulation therapy, which produces coordinated muscle contractions allowing people with paralysis to perform functional movements with rich sensory feedback; robot-assisted therapy, which uses robots to assist, resist, and guide movements with increased intensity while also reducing the physical burden on therapists; and brain–computer interfaces, which make it possible to verify the presence of motor-related brain activity during rehabilitation. Further, the book presents the combined use of these three technologies to illustrate some of the emerging approaches to the neurorehabilitation of voluntary movement. The authors share their practical experiences obtained during the development and clinical testing of functional electrical stimulation therapy controlled by a brain–computer interface as an intervention to restore reaching and grasping.

KEYWORDS

stroke, spinal cord injury, neurorehabilitation, voluntary movement, functional electrical stimulation, robot-assisted therapy, brain–computer interfaces

Contents

Acknowledgments

The authors wish to thank Dr. Milos R. Popovic for his guidance, unwavering support, and friendship, and Lazar I. Jovanovic, Aaron Marquis, and Dr. Chaim Katz for their invaluable role in creating and testing the technology described in Chapter 6.

CHAPTER 1

Stroke, Spinal Cord Injury, and Neurorehabilitation

For individuals who suffer from acquired neurological conditions secondary to trauma to the brain or spinal cord, rehabilitation offers them their best chance at regaining lost function and improving their level of independence and, thereby, their quality of life. This becomes even more relevant given that medical advances have contributed to a stark improvement in medical care, thereby reducing deaths associated with, for example, stroke and spinal cord injury (SCI). The American Heart and Stroke Association acknowledged that there has been a drastic decrease in the mortality rate in stroke, quoting it as one of the most significant achievements of the 20th century [1]. Similar trends have been observed for SCI as well. This is welcome news to the medical field. For the rehabilitation community, this has translated into increased pressure to conduct research and develop therapies that have high efficacy and are relatively easy to administer within the constraints of the healthcare system.

Keeping pace with the advances in other areas of medicine, rehabilitation care has also experienced a tremendous revolution over the past three decades. The starkest change being that the focus of care has realized a paradigm shift from care tailored towards improving function by compensation, to care delivered with the end goal of physiological function restoration. An overly simplified example would be to teach an individual who has suffered a left middle cerebral artery (MCA) infarct to use the left hand for all activities of daily living (ADLSs) although he/she might be right-handed dominant prior to stroke.

This chapter provides a brief background on the epidemiology, pathophysiology, and clinical presentation of two of the most debilitating and life-altering conditions: stroke and spinal cord injury. Subsequent chapters will cover advancements in the field of rehabilitation using various technologies specific to this population. We choose to focus on stroke as the World Health Organization has identified it as the second leading cause of death and the third leading cause of disability and has called upon a global response to tackle this [2]. The motivation to include SCI rehabilitation is the fact that, although numbers in terms of incidence and prevalence are low for this condition, it leaves a devastating impact on the individual, family and community, and there is a high economic impact on the healthcare system to care for individuals with SCI.

1.1   STROKE

1.1.1   DEFINITION AND EPIDEMIOLOGY

Stroke or cerebrovascular accident (CVA) is defined as a sudden non-convulsive focal neurologic deficit [3]. Stroke is the third leading cause of death in North America [4–6] and remains the main cause of morbidity and disability in survivors. Every year in the U.S. there are approximately 700,000 new stroke cases—roughly 600,000 ischemic lesions and 100,000 hemorrhages, intracerebral or subarachnoid—with 175,000 fatalities from these causes combined. In Canada, stroke remains a leading cause of adult disability, with over 400,000 people living with its effects [7]. By 2038, the number of Canadians living with the effects of stroke is expected to increase to between 654,000 and 726,000 [7].

Besides the physical deficits and consequences, stroke reigns a multi-fold impact, including financial consequences. According to a study published by Goeree et al., the Canadian government spends approximately 3% of its national healthcare expenditure on stroke, and with the aging population, this percentage is likely to increase in the near future [8]. Stroke-related costs in the U.S. came to nearly $46 billion between 2014 and 2015 [9]. This total includes the cost of health care services, medicines to treat stroke, and missed days of work. A review of economic studies shows a wide range of per-patient costs from $468–146,149 (U.S.), with few studies examining costs after hospital discharge. Of all stroke cases, 20% of patients will need medical care and rehabilitation after the CVA event [4–6].

1.1.2   TYPES, PATHOPHYSIOLOGY, AND CLINICAL PRESENTATION

Stroke can be classified as ischemic or hemorrhagic based on the underlying pathology. Ischemic stroke results from a brain vessel occlusion (by a thrombus or emboli) and blockage, and it accounts for 80% of all strokes [4–6]. Ischemic stroke is hence also referred to as a thrombotic or an embolic stroke. The resulting neurologic syndrome corresponds to the area of the brain that is supplied by the affected vessels. Hemorrhagic stroke occurs due to a blood vessel rupture (intracerebral or subarachnoid hemorrhage) and accounts for 20% of CVAs.

The presentation following stroke is closely aligned to the area of the brain affected. Among ischemic strokes, 84% of them occur due to involvement of the middle cerebral artery (MCA). The MCA is a major vessel supplying the lateral areas of the frontal, temporal, and parietal lobes, including the primary motor cortex and the sensory areas of the face, throat, hand, and arm, and, in the dominant hemisphere, the areas of speech. The other vessels involved might be the anterior or posterior cerebral artery. Most embolic strokes are sudden onset, and the deficits reach their peak almost immediately. On the other hand, hemorrhagic strokes evolve more slowly, over a period of minutes or hours and, occasionally, days. Another presentation of stroke is a Transient Ischemic Attack (TIA). Neurologically, a TIA is less devastating and is characterized by a rapid regression of a focal stroke syndrome that reverses itself entirely and dramatically over a period of minutes or up to one hour.

The clinical presentation post-stroke reflects both the location and the size of the infarct or hemorrhage. Hemiplegia stands as the most typical sign of cerebrovascular diseases. However, there is a myriad of associated deficits. Table 1.1 provides an overview of the signs and symptoms encountered based on the anatomical location of the injury [10].

Table 1.1: Clinical presentation of stroke		
Clinical deficit	**Injury location**	**Lobe affected**
Unilateral weakness (affecting face, arm, and leg)	Internal capsule Corticospinal tract	Variable
Unilateral weakness (affecting face and arm > leg)	Lateral Motor cortex	Frontal lobe
Unilateral hemiparesis (affecting leg and trunk)	Medial Motor cortex	Frontal lobe
Aphasia	Broca's area or Wernicke's area	Dominant hemisphere injury: Broca's area: frontal lobe, Wernicke's area: temporal lobe
Visual deficits	Frontal eye fields	Frontal lobe
Sensory deficits	Nondominant parietal lobe	Nondominant Hemisphere injury: Parietal lobe
Visual field deficit	Occipital cortex (PCA (posterior cerebral artery)), Optic radiations (MCA (middle cerebral artery))	PCA: occipital lobe, MCA: parietal lobe (inferior quadrantanopsia), temporal lobe (superior quadrantanopsia). Large MCA strokes cause hemianopsia

1.2   SPINAL CORD INJURY

1.2.1   DEFINITION AND EPIDEMIOLOGY

Based on etiology, SCI can be classified into Traumatic Spinal Cord Injury (tSCI) and Non-Traumatic Spinal Cord Injury (ntSCI). tSCI occurs when a sudden, traumatic impact on the spine compromises the spinal column or when there is a fracture or dislocation of vertebrae, subsequently causing damage or compression to the spinal cord [11]. ntSCI refers to compression of the spinal cord related to a tumor, infection, or degeneration of the spinal column. The mechanism, in this case, is a gradual compression (slow spinal cord injury) of the cord due to an expanding lesion or progressive degeneration causing increased compression over time [12–14]. The annual incidence of SCI is approximately 54 cases per one million people in the U.S., or about 17,730 new SCI cases each year [15]. The estimated number of people living with SCI in the U.S. is approximately 291,000, ranging from 249,000–363,000 individuals [16]. The average age of SCI has increased recently from 29 years during the 1970s to 43 years [17]. About 78% of new SCI cases are male [17]. Vehicle crashes are the most recent leading cause of injury, closely followed by falls. Incomplete tetraplegia (described below) is the most frequent neurological category, with 47% of traumatic SCI's falling in this category [17]. The frequency of incomplete and complete paraplegia is the same. Less than 1% of people experience complete neurological recovery by the time of hospital discharge. In Canada, the estimated initial incidence of traumatic spinal cord injury (tSCI) is 1,785 cases per year, and the discharge incidence is 1,389 (41 per million) [18]. The estimated discharge incidence for non-traumatic spinal cord injury (ntSCI) is 2,286 cases (68 per million). The prevalence of SCI in Canada is estimated to be 85,556 persons (51% tSCI and 49% ntSCI) [18].

1.2.2   TYPES OF SCI, PATHOPHYSIOLOGY, AND CLINICAL PRESENTATION

SCIs are classified based on the level and severity of the injury. Accordingly, the two main categories of spinal cord injury are Complete and Incomplete injuries. This is more clearly demarcated in injuries of traumatic origin. A complete SCI refers to loss of all motor and sensory function including, sacral segment S4-S5. In comparison, an incomplete injury presents with some preservation of sensory or motor function below the neurological level, including the sacral segments S4-S5. The commonly used scale to classify an SCI based on the severity of the injury is the American Spinal Cord Association Impairment Scale (AIS) [19]. Spinal cord injury categorization also takes into consideration the level of injury, i.e., cervical, thoracic, lumbar, or cauda equina injury. Typically, higher levels of injury are associated with greater deficits. Injury at the cervical level results in Tetraplegia (i.e., all four limbs, including the trunk, are involved), whereas injury below the cervical level results in Paraplegia (i.e., involvement of the trunk and the lower limbs).

Ultimately, irrespective of the level and severity of injury, SCI has a presentation of motor and sensory deficits coupled with functional loss. In other words, all gradations of SCI manifest with a degree of paralysis or paresis as a result of an upper motor neuron lesion. Over and above the motor and sensory deficits, there is a myriad of other clinical manifestations, including an initial period of spinal shock, impaired temperature control, respiratory impairment, spasticity, bladder and bowel dysfunction, sexual dysfunction, and secondary complications like pressure sores, autonomic dysreflexia, postural hypotension, contractures, deep venous thrombosis, osteoporosis, and traumatic pain. For additional details please refer to *Principles of Neurology* by Adams et al., [3].

1.3   REHABILITATION FOLLOWING STROKE AND SPINAL CORD INJURY

Rehabilitation commences once a patient is medically stable and initially may focus on preventing secondary complications. The general objectives of rehabilitation following a brain or spinal cord injury include (1) maximizing independence in activities of daily living, including self-care, and mobility; (2) assist with the acceptance of a new lifestyle concerning modifications with recreational activities and housing options; and (3) finally and most importantly assist with reintegration into society [20].

Rehabilitation for patients with stroke and SCI has evolved over the years and is undergoing a paradigm shift. In the past, rehabilitation specialists focused on how best and how quickly individuals with deficits could go back to living an independent life, even if that meant teaching compensatory strategies. Levin et al. have elegantly defined compensatory changes versus recovery at various levels of movement execution [21]. At an activity level, they define recovery as follows: "Recovery at the Activity level requires that the task is performed using the same end effectors and joints in the same movement patterns typically used by nondisabled individuals." Compensatory movement is defined by the same group as "compensation at this level often takes the form of substitution and would be noted if the patient were able to accomplish the task using alternate joints or end effectors" [21]. The authors provide a simple example to illustrate the difference between the two strategies wherein using a compensatory strategy to open a jar of food would involve using 1 unaffected extremity and the mouth versus using both hands [21]. With an increasing understanding of neuroplasticity, goals have shifted toward neuromuscular re-education and recovery of lost function. During the acute and subacute phases of treatment, rehabilitation strategies typically focus on preventing secondary complications, promoting neurorecovery and maximizing function [20]. In the chronic phase, there is a shift to compensatory or assistive approaches. The optimal management strategies for patients with stroke and SCI are difficult to define due to the wide variations in presentation, lack of standardization of interventions, therapeutic doses and outcome measures, heterogeneous populations, superimposed spontaneous recovery, and occasionally a lack of clear best practice guideline [22]. Nevertheless, clinicians and researchers have systematically written clinical care guidelines for front-line clinicians to follow for both stroke [23] and SCI [24]. We encourage readers interested in current clinical best practices to refer to these guidelines.

1.4   NEUROPLASTICITY AND NEUROPLASTICITY-BASED INTERVENTIONS

The brain's capacity to function is dependent on the number of functional neuronal connections, and new learning is determined by the capacity of the brain to form new functional connections. Plastic phenomena are at the basis of learning and damage repair. Cortical maps can be modified by sensory input, experience, and learning. Cortical representation areas probably undergo continuous transient changes during routine life experiences in response to repeated stimuli, movement patterns, and cognitive tasks. Cramer et al. defined neuroplasticity as "the ability of the nervous system to respond to intrinsic or extrinsic stimuli by reorganizing its structure, function and connections" [25]. Valuable insights into the neurophysiological mechanisms mediating neural plasticity have emerged from the introduction of non-invasive techniques in studies on humans, including positron emission tomography (PET), transcranial magnetic stimulation (TMS), and functional magnetic resonance imagery (fMRI).

Concerning upper motor neuron injuries, neuroplasticity has been extensively studied for motor recovery following stroke [25]. Motor recovery following stroke is facilitated by neuroplasticity at various levels within the nervous system. Neural synapses and neuronal circuits can change because of activity. The brain, and particularly the cortex, has the capacity to change its structure and consequently function during learning or in response to exposure to enriched environments. Similar plasticity mechanisms have been observed across divergent forms of central nervous system injury, including SCI suggesting that plasticity, as with development, uses a limited repertoire of events across numerous contexts. Another widely discussed concept related to neuroplasticity is the timing of heightened plasticity, i.e., generally speaking, intense rehabilitation following the principles of neuroplasticity offered early on after injury (within the first 6–12 months post-injury) results in better outcomes. However, there is also emerging evidence related to the benefits of intense task-specific training during chronic phases post-injury.

Neurorehabilitation techniques exploit this plastic nature of the CNS. Kleim and Jones have identified ten principles of experience-dependent plasticity that one might want to integrate during rehabilitation procedures [26]. These principles include the following.

1. **Use It or Lose It:** Failure to drive specific brain functions can lead to functional degradation.

2. **Use It and Improve It:** Training that drives a specific brain function can lead to an enhancement of that function.

3. **Specificity:** The nature of the training experience dictates the nature of the plasticity.

4. **Repetition Matters:** Induction of plasticity requires sufficient repetition.

5. **Intensity Matters:** Induction of plasticity requires sufficient training intensity.

6. **Time Matters:** Different forms of plasticity occur at different times during training.

7. **Salience Matters:** The training experience must be sufficiently salient to induce plasticity.

8. **Age Matters:** Training-induced plasticity occurs more readily in younger brains.

9. **Transference:** Plasticity in response to one training experience can enhance the acquisition of similar behaviors.

10. **Interference:** Plasticity in response to one experience can interfere with the acquisition of other behaviors.

We encourage readers to refer to Kleim and Jones et al. for a detailed understanding of these principles [26].

Whereas neuroplasticity occurs at various levels within the CNS, one critical component to enhance the effects of neuroplasticity is to engage the entire efferent-afferent loop in a functional manner, repeatedly. Interventions that are developed on the premise of neuroplasticity and that are either already in use in clinical practice or rapidly gaining attention from clinicians include constraint-induced movement therapy, activity-based therapy (ABT), functional electrical stimulation therapy (FEST), and robot-assisted therapy. All of these interventions aim to engage the entire neural circuitry during the performance of repetitive task-specific activities. Below, we introduce FES and robot-assisted therapy, both also treated in detail in later chapters.

1.4.1   FUNCTIONAL ELECTRICAL STIMULATION THERAPY

As mentioned in the previous section, one important example of neuroplasticity-based interventions that have become an important tool for rehabilitation of voluntary movement after paralysis is functional electrical stimulation therapy or FEST [27]. In this rehabilitation approach, patients practice different tasks with their impaired limbs. The tasks are selected according to each individual's goals and abilities and are adjusted throughout the therapy to accommodate any changes in the motor abilities. At every moment, the practiced movement is assisted simultaneously by a therapist and through muscle contractions produced by electrical pulses [28, 29]. The muscles are selected carefully to ensure that they produce the movement being practiced when they contract in response to the electrical stimulation. In a typical FEST session, patients can practice several tasks, and each one is repeated multiple times. A FEST intervention can last multiple sessions, delivered several times per week over multiple months [30].

The electrical stimulation is delivered with one or multiple stimulation channels, depending on the movement that is practiced. Each channel, consisting of two leads (i.e., anode and cathode), can typically stimulate one muscle [27]. Accordingly, multiple channels make it possible to produce complex movements (for example, reaching and grasping). The stimulation is often delivered with electrodes placed on the skin, although there are also invasive methods.

FEST has been successful in improving voluntary movement in individuals who have had a stroke or an SCI, including individuals with very high levels of impairment (i.e., the ability to move is severely limited or completely lost), who are often unable to participate in movement rehabilitation. It is believed that the efficacy of FEST lies in its ability to facilitate neuroplastic changes, resulting from the simultaneous presence of a descending motor command (i.e., efferent activity, produced when a movement is attempted) and the corresponding somatosensory feedback (i.e., afferent activity) resulting from the electrically-generated muscle contractions [31]. FEST is presented in detail in Chapter 2.

1.4.2   ROBOT-ASSISTED THERAPY

Another important development in rehabilitation has been the integration of robotic systems, giving rise to what is often referred to as robot-assisted therapy. Robots can assist movements of the upper and lower limbs. For example, they can be incorporated into the rehabilitation of reaching and grasping, which involve the shoulder, elbow, and hand, and in treatments focused on restoring the ability to walk, requiring the activity of the hips, knees, and ankles. Movement with a robot can be achieved by applying force to the distal end of a limb (e.g., a hand or a foot), or these machines can take the form of an exoskeleton, applying force around multiple joints.

Robotic systems have multiple modes of operation to interact with a paretic or paralyzed limb. In the context of rehabilitation, they can assist or resist movements and constrain trajectories of a limb in motion. They can also provide anti-gravity assistance and may also offer the possibility of facilitating movements on one hand or leg, suitable for individuals with hemiplegia following stroke, or on limbs on both sides of the body, more typically required while treating individuals with SCI.

In addition to producing movements, robots can also use their built-in sensors to record kinematic and kinetic information and provide quantitative information throughout an intervention. These data can provide an additional dimension to guide treatment and can serve as an important research tool. Using the same sensing systems, some robots can also integrate interactive visual feedback, presented as a video game that can provide guidance and user engagement.

Robot-assisted therapy has the potential to enhance rehabilitation by increasing the intensity of therapy. In the context of rehabilitation, intensity can be defined in multiple ways, including the number of times a task is repeated during practice or the number of sessions held over a period (e.g., a week). More importantly, both of these definitions are listed in the principles of experience-dependent plasticity (listed in the previous section) as factors that promote neuroplasticity. The assistance provided by robots can also reduce the physical burden for the therapist, enhancing the potential of these systems to deliver rehabilitation (e.g., by allowing to increase the number of patients that a therapist can serve). Readers interested in robot-assisted therapy are directed to Chapter 3.

1.5   CHALLENGES IN REHABILITATION OF VOLUNTARY MOVEMENT AFTER STROKE AND SCI

Despite impressive advances in neurorehabilitation, there is still a need to improve rehabilitation strategies [32]. It has been reported that over 40% of survivors require assistance while performing activities of daily living years after stroke [33]. Many stroke survivors of all ages experience lifelong impairments after rehabilitation [34–36]. Similarly, the effects of SCI on motor function are present for the remainder of a person's life. In addition, the number of stroke cases is expected to grow as the population ages [37].

One of the most recent innovations in rehabilitation after stroke is the use of brain–computer interfaces (BCI). This technology can detect a person's intention to perform a movement from a person's brain, providing a unique opportunity to ensure that efferent activity is present at the highest (i.e., cortical) level, even in individuals where the ability to move is significantly impaired or lost. This capability directly supports some of the neuroplasticity-guided neurorehabilitation strategies. As such, the last decade has seen an exponentially growing interest in exploring BCI technology as a tool to promote, and perhaps enhance, neuroplasticity leading to the recovery of the ability to move voluntarily. Chapter 4 onward, we describe the use of BCIs as a tool for the neurorehabilitation of voluntary movement after stroke and SCI.

CHAPTER 2

Functional Electrical Stimulation Therapy: A Closer Look

2.1 HISTORY OF FUNCTIONAL ELECTRICAL STIMULATION THERAPY (FEST)

The use of an electrical current to manipulate human tissues dates back to several decades. Ever since Galvani's demonstration in 1792 that living tissue reacts to electrical currents, the use of electricity in living tissues has preoccupied the minds of many scientists [38]. Among its many uses in medicine, one is in the form of functional electrical stimulation therapy (FEST), which may be defined as the stimulation for rehabilitation purposes of neuromuscular units normally under voluntary control [38]. Functional electrical stimulation (FES) can also be defined as a systematic and co-ordinated application of electrical current to excitable tissues to supplement or replace function that is lost in neurologically impaired individuals. Both sensory and motor function can be restored with FES [39]. Auditory and visual neuroprostheses are examples of FES used to restore sensory system functions. Neuroprostheses used for grasping and walking are examples of FES systems for replacing motor function; a functional limb movement can be produced by properly coordinating several electrically activated muscles [39]. The concept is to provide functional restoration through electrical activation of intact lower motor neurons using electrodes placed on or near the innervating nerve fibers. Appropriate electrical stimuli can elicit action potentials in the innervating axons, and the strength of the resultant muscle contraction can be regulated by modulating the stimulus parameters.

FES was first applied in the 1960's for retraining gait after stroke [40]. Liberson and his co-workers stimulated the dorsi flexors of the ankle joint in hemiplegics synchronously with gait to treat drop foot [40]. This successful application gave rise to numerous research projects, which extended to fundamentally new areas and to the application of a new technology to rehabilitation. The paradigm was to use it as an orthosis so that the patient would don it when needed and take it off when the task no longer needed to be done. Around the same time, Liberson, Long, and Masciarelli [41] developed the first FES hand splint which was designed as a one-channel device enabling the patient to control hand opening combined with wrist extension.

Even back then the technology was realized to have two potential applications [38]:

1. as a neuroprosthesis for replacing lost function and

2. as a short-term therapeutic device for retraining lost function.

Since the inventions of Liberson and colleagues, FES technology has been researched and refined significantly. It has gained traction among rehabilitation specialists for its ability to restore function and thereby improve quality of life (QOL) of individuals living with the sequelae of neurological injuries or diseases. Neuroprostheses for replacing lost function have been mostly in the form of implanted devices. This option is attractive in individuals where chances of recovery of lost function are minimal. The prosthesis can be implemented in a variety of ways, such as through stimulation of the spinal cord [42], nerve [43], or muscle [44]. This application has not gained as much attention from the rehabilitation professionals given the invasive nature of the treatment and several other documented limitations. On the other hand, short-term therapeutic FES applications for restoring upper and lower extremity motor function following stroke or SCI (i.e., FEST) are becoming a part of the rehabilitation tool kit in many rehabilitation hospitals and clinics.

2.2   TRANSCUTANEOUS FES SYSTEM

A transcutaneous FES stimulator is non-invasive, and its components can be divided into hardware and software.

2.2.1   FES HARDWARE

The stimulator hardware consists of three components: (a) the electrical stimulator, (b) the electrodes used to deliver the stimulation, and (c) the user interface. The electrical stimulator is typically a multi-channel stimulator and may have anywhere from 2–8 stimulation channels; a higher number of channels allows for simultaneous stimulation of multiple muscles or muscle groups. The electrodes used for transcutaneous stimulation are typically self-adhesive carbon electrodes. Alternatively, surface electrodes that can be secured using adhesive tape may also be used. Different simulators have different types of user interfaces. Push buttons, electromyography (EMG)/biofeedback sensors, and sliding potentiometers are examples of the devices that have been used to control neuroprosthetic systems.

2.2.2   FES SOFTWARE

The software of the FES system allows one to specify/alter all the stimulation parameters. Based on individual stimulators, the parameters that can be changed are stimulation frequency, the intensity range (i.e., minimum and maximum), pulse width (i.e., duration), ramp-up time, synchronization with other stimulations, number of targeted repetitions, and interactions with the user. The frequency of stimulation (i.e., the rate at which stimulation pulses are delivered) affects the strength and quality of the muscle contraction. Producing a tetanic contraction requires a minimum stimulation frequency of 16–20 Hz [45]. However, a pulse frequency of 40 Hz is often needed. Higher pulse frequencies generate stronger tetanic contractions, but they can also result in faster muscle fatigue [27]. The pulse amplitude/intensity refers to the magnitude of the stimulation. It directly affects the type of nerve fibers that respond to the stimulation, with large fibers near the stimulation electrode being recruited first [27]. The pulse duration (pulse width) refers to the amount of time the stimulation pulse is present. The ramp-up function augments the electrical charge successively by prolonging the pulse width from a selected minimum to a chosen maximum value.

Typically, prior to a FEST session the sensory, motor, functional and maximum intensity are assessed at 40 Hz frequency and 300 µs pulse duration or with the parameters of stimulation that will be used during the therapy session.

2.3   REHABILITATION IN STROKE AND USE OF FEST

The Canadian Stroke Best Practice Recommendations state that "the rehabilitation process offers people with stroke their best opportunity for optimal recovery" [46]. Rehabilitation can be defined as a "progressive, dynamic, goal-oriented process aimed at enabling a person with impairment to achieve their pre-stroke level of physical and social functioning and can commence as soon as the patient is medically stable" [46]. There is strong evidence that mortality rates and likelihood of institutional care are reduced with organized, inter-professional care delivered in inpatient or outpatient and community-based settings. This rehabilitation model also enhances recovery and independence [47–50]; rehabilitation can be further defined by factors such as the initial severity of the stroke, individual progress, and availability to participate in therapy [46]. The pressure on providers of post-acute rehabilitation to deliver clinically effective care will only intensify as cost-containment efforts and requirements for documenting the quality-of-care increase. Systematically acquired evidence can guide rehabilitation services toward higher-quality, effective, and cost-efficient care; however, the evidence for certain conditions and rehabilitation settings is better developed than for others. More clearly, delineating the evidence of effectiveness will help determine whether certain post-acute rehabilitation services produce better outcomes than alternatives. Subsequently, policy-makers, health care administrators, and clinicians might be better informed for making decisions about providing rehabilitation services that help patients attain functional autonomy and an improved quality of life. The same recommendations state that "all patients with acute stroke should be assessed to determine the severity of stroke and early rehabilitation needs" [51]. Further, they state that initial screening and assessment should ideally be commenced within 48 h after a patient is admitted to the hospital and that they should receive rehabilitation as soon as possible.

While there are many modalities that are used to retrain function, we continue to seek for the best practices and treatment strategies that can more effectively tackle the deficits and restore independence. In that quest, functional electrical stimulation has received a great deal of attention.

FES has been researched as an adjunct to existing therapies as well as in combination with newer rehabilitation modalities.

2.3.1   LOWER LIMB FUNCTION IN STROKE AND ROLE OF FEST

In the lower extremity, FES has been used to decrease spasticity and to improve muscle strength as well as walking ability. However, almost always the end goal of FEST to the lower limb in stroke is to improve gait. Canadian rehabilitation guidelines recommend that patients should participate in training that is meaningful, engaging, progressively adaptive, intensive, task-specific, and goal-oriented in an effort to improve transfer skills and mobility.

FES has long been used to treat gait deficits in stroke. In the most basic form, transcutaneous 2-channel FES devices are used to stimulate ankle dorsiflexors in an attempt to correct drop foot. However, these single or dual channel FES devices stimulate ankle dorsiflexion alone, thereby failing to improve gait deficits at the knee and hip. Peroneal nerve stimulators (PNS) target dorsiflexion movement during swing phase whereas use of multi-channel stimulators allows for addressing deficits around the hip, knee, and ankle. According to literature, PNS stimulators have shown limited efficacy in improving gait. In a randomized control trial conducted by Shefflar et al., the authors compared the motor relearning effect of a surface PNS versus usual care on lower limb motor impairment, activity limitation, and quality of life among chronic stroke survivors and found that PNS was not superior to usual care in individuals with chronic stroke [52]. In a recent review conducted by Maira Jaqueline da Cunha et al., the authors reported on a meta-analysis that revealed a low quality of evidence for positive effects of FES applied on the paretic peroneal nerve on gait speed when combined with physiotherapy [53].

In contrast, studies that have looked at the effect of multi-channel stimulation (four-channel stimulation) have shown positive effects. A study led by Stanic [54] has shown that multichannel FEST given 10–60 min per day, 3 times per week for 1 month, improved the gait performance of hemiplegic subjects. Bogataj and his colleagues [55] have applied multichannel FES to activate the lower limb muscles of 20 chronic hemiplegics. After a treatment lasting between 1 and 3 weeks, with daily sessions delivered 5 days per week, non-ambulatory participants were able to walk again. Thus, positive effects of multichannel and conventional FEST on the affected lower extremity have been demonstrated in several studies [54–57]. Recent studies are now looking at using FEST in combination with not only conventional rehabilitation strategies but also robotics and virtual therapy. Another newer concept is FastFES in which patients walk on a treadmill as fast as they can with assistance delivered with electrical stimulation to the ankle, i.e., ankle plantar and dorsiflexors [58, 59]. In conclusion, therapeutic effects of FEST have been demonstrated at the body function and activity levels when used as a training modality by various RCTs. However, when compared to matched treatments that did not incorporate FES, no definite conclusions can be drawn regarding the superiority of FEST [60].

Commercially available surface FES devices, approved by the U.S. Food & Drug Administration (FDA), are shown next.

Odstock: The Odstock dropped foot stimulator (ODFS) is one of the FES systems with the most widespread clinical use in the United Kingdom, where FES is used extensively. The device consists of a single-channel foot switch-triggered stimulator designed to elicit dorsiflexion of the foot by stimulation of the common peroneal nerve. The stimulator uses surface stimulation electrodes, and the timing of the stimulation is adjusted to prevent eliciting the stretch reflex as well as to prevent foot slap. Several studies have reported a positive training effect with the use of ODFS [61, 62], including significant improvements in walking speed and reduction in walking effort in individuals who had a stroke [63] and a motor-incomplete SCI [64]. The latest version of the device is the ODFS Pace® (Figure 2.1), commercialized by Odstock Medical Ltd, U.K.

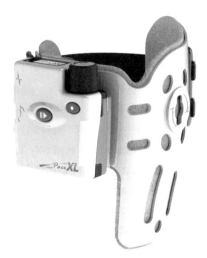

Figure 2.1: ODFS Pace® with a leg cuff (https://www.odstockmedical.com).

NESS L300: The NESS L300 delivers electrical pulses to the common peroneal nerve throughout the swing phase of gait, resulting in ankle dorsiflexion to prevent foot drop (Figure 2.2). The system consists of three main components that communicate via radio frequency signals. (1) A hybrid orthosis with an integrated stimulation unit and electrodes. One electrode is located over the common peroneal nerve, posterior and distal to the fibular head, and a second electrode is located over the tibialis anterior muscle to achieve dorsiflexion with slight eversion. The movement may be further adjusted by modifying the position of the electrodes during the fitting process. (2) A gait sensor that detects the force under the foot, using a force-sensitive resistor. (3) A miniature control unit. The orthosis ensures contact between the user's limb and the electrodes as well as reproducibility of electrode placement [65].

Figure 2.2: **NESS L300** (https://www.bioness.com).

WalkAide foot drop stimulator: The WalkAide consists of a single-channel stimulator mounted on a cuff that fits around the upper part of the shank and calf (Figure 2.3). Two round hydrogel electrodes (diameter 3.2 cm) are attached to the inside of the cuff with Velcro. The active electrode is typically positioned over the common peroneal nerve just distal and dorsal to the head of the fibula, whereas the indifferent electrode is placed over the tibialis anterior muscle belly. Precise positioning of the electrodes usually results in adequate ankle movement balancing dorsiflexion with eversion. To synchronize the stimulation with the swing phase of the gait cycle, the WalkAide uses a tilt sensor or a heel sensor. Studies have shown a positive effect of Walk Aide on walking speed in individuals with stroke [66].

Figure 2.3: **WalkAide** (https://acplus.com/walkaide).

2.3.2   UPPER LIMB FUNCTION IN STROKE AND ROLE OF FEST

The therapeutic effects of FEST for restoring upper extremity function are discussed in literature as early as in 1978. In a study published by Vodovnik et al., the authors talk about the reappearance of functional movements during electrical stimulation as one the major effect of FES [67]. However, up until 1992, the use of surface FES was still restricted to stimulation of muscles of the forearm to stimulate hand opening although a great deal of specificity was obtained using surface stimulation with high-resolution surface electrodes [68]. The next decade saw exponential work in this field and by 2002 there were various FES grasping systems available, some commercially, and some as research devices. In the last two decades, FES has received exponential attention from clinical and biomedical engineers resulting in significant refinement of the technique and a much more sophisticated application. Therapeutic FES (i.e., FEST) developed over the past 10–15 years is based on the premise of activity-based therapy wherein it aims to accomplish repetitive task-specific practice. FEST has been combined with various other rehabilitation modalities like mirror therapy, virtual therapy, robotic-assisted therapy, tele-supervised therapy, etc. There are a few review articles published in literature that are related to the efficacy of FES in improving activities of daily living. These articles provide conflicting evidence in terms of efficacy, but most agree that FEST applied during acute rehabilitation process results in better outcomes as compared to conventional therapies.

The most commonly discussed systems in literature include the Freehand system, NEC FES Mate system, NESS H200, Bionic Glove, ETHZ-ParaCare neuroprosthesis, the systems developed by Rebersek and Vodovnik, the Belgrade Grasping–Reaching System, and Compex Motion stimulator. All of these systems were primarily designed and researched in individuals with SCI except for the NESS H200 and the Compex Motion stimulator. Xcite is another multi-channel FES stimulator. Xcite is a portable system with easy to set up pre-programmed activity libraries, that deliver sequenced stimulation through up to 12 stimulation channels [51] and is used both in rehabilitation after stroke and SCI. The Compex Motion simulator is a four-channel surface FES device that is fully programmable and allows for designing stimulation protocols based on individual patient abilities. This stimulator has been extensively used in research with positive outcomes for even chronic and severely impaired stroke patients [69, 70]; however, this is a research-based stimulator and is no longer widely available.

To date, the only surface FES devices that are FDA approved and designed for use in individuals with stroke are the Handmaster or the Neuromuscular Electrical Stimulation System (NESS) H200 and the MyndMove.

> **NESS H200:** The NESS H200 was invented by Roger Nathan and his group in Ben-Gurion University, Israel. It is a three-channel surface FES device consisting of two parts: the stimulator and the forearm splint. Control of the device is via a user activated push button (Figure 2.4). The stimulator generates electrical impulses that are delivered to the target muscles by hydrated sponge surface electrodes, held in position by the forearm splint. This device is programmed to perform three exercise modes and two functional modes. The exercise mode provides repetitive stimulation to a group of muscles, thereby helping the user to increase muscle strength. Functional modes, on the other hand, help the user in performing key grip and palmar grasp and, hence, help them in performing their ADL [71].

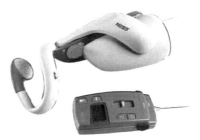

Figure 2.4: NESS H200.

> **MyndMove:** The MyndMove stimulator is a FES device that is non-invasive and uses short, low-energy electrical pulses to induce muscle contractions (Figure 2.5). The device offers a full range of protocols designed specifically for MyndMove therapy to address proximal and distal impairments of the upper extremity. Up to eight muscle groups can be stimulated during a protocol. Thirty protocols, of which 17 are for use in stroke and 13 for use in SCI, provide a full range of reaching and grasping movements that are broken down into sub-movements that can be initiated by the therapist or patient using hand or foot switches. Electrode positioning for different protocols is facilitated by anatomical illustrations on the device that indicate the desired location for electrode placement on specific muscle groups. Amplitude for

electrical stimulation of a specific muscle group is adjusted up to a pre-set maximum for all protocols and for each user prior to every treatment session [72].

Figure 2.5: MyndMove.

In our laboratory, we have applied FEST to retrain reaching and grasping function in individuals with both acute and chronic stroke using the four-channel surface Compex Motion stimulator. We have typically used two channels to facilitate proximal shoulder and elbow movement and two channels to assist with hand opening. Individuals with stroke generally present with a flexor synergy of the arm with a strong component of finger flexion and thus hand closing is generally not elicited using FES in this patient population. Although we have not done a systematic study related to optimal dosing, we have found that it takes a minimum of 20 FEST sessions to see any change on functional outcome assessments, and 40 sessions to see an improvement in functional status.

2.4   REHABILITATION AFTER SPINAL CORD INJURY (SCI) AND USE OF FEST

Rehabilitation after a SCI is a lengthy process aimed at restoring as much independence as possible given the level and extent of injury. Individuals with injury in the cervical region face paralysis of bilateral upper and lower extremities, although the extent of involvement of all four limbs could be different where one or more of them might be more severely affected compared to the rest. In individuals with cervical injury, restoring upper extremity function promotes independence with day-to-day activities such as eating, grooming, bathing, and toileting, as well as mobility related activities like transfers and wheelchair mobility [73]. Individuals with SCI may require full assistance from a caregiver or may be partially functional in activities of daily living, and social, recreational, and work-related activities [74]. Accordingly, restoration of upper limb function was rated above control of bladder and bowel function, spasticity, pain, and sexual function in individuals that have experienced a SCI [75]. With advances in medical care, the life expectancy post-SCI has increased significantly. Given that rehabilitation remains the mainstay of treatment for restoring lost function, researchers are working hard to develop rehabilitation techniques that can optimize recovery. Among the various modalities that are being researched, FEST has also gained significant attention in this domain. Both invasive and non-invasive FES stimulators are researched to restore function. The invasive stimulators look at stimulating the spinal cord or target nerves and muscles using epidural, epimysial or percutaneous electrodes. However, to date, most of these remain research tools with very little clinical uptake of the technology. In contrast to this, transcutaneous stimulators have found a higher acceptance as both an orthosis as well as a neuromodulation therapeutic device.

The overall management of individuals with tetraplegia often emphasizes the care and rehabilitation of the upper limbs, given their importance in maintaining and maximizing independence. This can include extensively practicing movements associated with ADLs (i.e., functional strength training), the use of orthosis to provide support and stability, and surgery. Functional treatment of the upper limb typically starts with conservative methods followed by FES and surgical interventions, if appropriate [76]. Although lower limb rehabilitation after SCI has focused on restoration of gait, regardless of whether or not functional walking is possible, these interventions help combat secondary complications by maintaining cardiovascular, muscle, bone, and skin health. In turn, the mitigation of these complications may translate into a decreased economic burden and promote active participation in society [51].

Conventional treatment strategies used for retraining function following SCI are heavily dictated by the level and completeness of injury. Functional goals are likewise established based on recovery potential. In complete SCI, the focus of rehabilitation is on teaching compensatory strategies to assist individuals regain as much self-independence as possible and mitigate the complications secondary to lack of mobility. In these patients, implanted FES orthosis that replace lost function have gained significant attention, but given the caveat around these, for the most part these treatments still remain research based. On the other hand, in incomplete injuries the conventional rehabilitation tool kit is comprised of therapies that help to increase strength and function below the level of lesion. This tool kit in the most traditional form consists of (a) muscle facilitation exercises emphasizing the neurodevelopmental treatment approach; (b) task-specific, repetitive functional training; (c) strengthening and motor control training using resistance to available arm or leg motion to increase strength; (d) stretching exercises; (e) electrical stimulation applied primarily for muscle strengthening; (f) practice of activities of daily living (ADLs) including self-care where the upper extremities are used as appropriate or gait training within parallel bars or with use of walking aids as appropriate; and (g) caregiver training. This might be complemented with treadmill training, FES cycling, robotic therapy, virtual therapy, haptics, etc. While we have made significant gains in terms of improving outcome following SCI, there is still a lot more to do to further these gains. Transcutaneous FES is being used in clinical care and continues to be explored and refined to maximize its benefits post SCI.

2.4.1 LOWER LIMB FUNCTION IN SPINAL CORD INJURY AND ROLE OF FEST

The goals of lower extremity rehabilitation vary based on level on injury. For a person with cervical injury, lower extremity rehabilitation may mainly be in the form of range of motion exercises to prevent muscle tightness that may interfere with wheelchair transfers, and to prevent development of pressure ulcers. Some patients in this category may train for physiological standing using orthosis and implantable or surface FES devices with a standing frame. The goal of such treatment might once again be prevention of secondary complications related to lack of mobility. In individuals with thoracic injury, the goals might be similar as above, especially for those with higher thoracic injuries. For lower thoracic injuries, these goals might extend to functional standing to perform short duration day-to-day tasks, and some might ambulate indoors with use of orthosis and walking aids or use of FES with walking aids. Standing and over ground ambulation training are important components of conventional rehabilitation using various bracing and assistive devices [77, 78], especially in lower-level injuries. For individuals who train to regain independent walking, there is an increasing emphasis on providing task-specific training of functional movements with the help of body weight support treadmills or overhead harnesses. There have also been exciting advances in technology applications for facilitating or augmenting gait rehabilitation strategies, using robotic devices for treadmill gait retraining [79, 80], introduction of microstimulators for activating paralyzed muscles [81] application of epidural spinal cord stimulation in combination with intensive therapy [82], and use of transcutaneous functional electrical stimulation in combination with the above mentioned conventional therapies or technological therapies with the goal of restoring lost function [30]. Studies combining FES with other technologies conducted in a laboratory or clinic setting have been predominantly supportive of a training effect [83, 84].

Some of the FES devices that use surface stimulation for lower extremity rehabilitation in SCI include the following.

Hybrid FES system: The system was developed by Andrews et al., and consists of sensors on the handles of crutches, a spinal brace, or on the ankle–foot portion of an ankle-foot orthosis (AFO). Goniometers and force sensitive resistors are used to determine the gait phase and, accordingly, time the stimulation [85].

Hybrid Assistive System: Hybrid assistive system (HAS) represents a combination of FES and a self-fitting modular orthosis controlled by artificial reflexes. It is an external neuroprosthesis for gait restoration in severely handicapped individuals [86].

Reciprocating gait orthosis (RGO) with FES: The FES used in combination with the RGO uses the Loewenstein modified Mark 1 stimulator. The stimulator contains 3 different operating programs addressed for exercising, standing, and walking. Two external switches provide a manual command. These switches are attached to each of the walker handles and serve for stand-up, sit-down and ambulation programs. The small size stimulator can be attached either to the subject's belt or to the walker itself [87].

Parastep: The system is composed of a six-channel, microprocessor-controlled stimulator, surface electrodes, and a modified walking frame designed to enable an individual with paraplegia to stand and ambulate short distances. It has been emphasized that the system is offered "as an alternative, not a substitute, for the wheelchair" [88].

Compex Motion stimulator: Compex Motion stimulator used for retraining lower extremity function uses surface self-adhesive stimulation electrodes secured to the lower extremity muscles (i.e., quadriceps, hamstrings, ankle dorsiflexors and ankle plantarflexors) [30]. Often two stimulators are used so that bilateral lower extremities can be stimulated (Figure 2.6). The electrodes are typically placed on the subject's skin corresponding to the muscles targeted with FES. Stimulus signals are balanced, biphasic, and pulse-width modulated with constant current regulation. Pulse amplitudes from 8–125 mA are used (subject and muscle dependent), and pulse-widths from 0–300 μs are used to modulate the stimulation intensity depending on the gait phase. Pulse frequency of 40 Hz is typically used [30].

Figure 2.6: Compex Motion Stimulator.

2.4.2 UPPER LIMB FUNCTION IN SPINAL CORD INJURY AND ROLE OF FEST

Today, there is little consensus on the management of upper limb in individuals with tetraplegia despite the existence of clinical guidelines (Consortium for Spinal Cord Medicine [265]; Consortium for Spinal Cord Medicine [266]). This is in part due to the variation in available function after

SCI [89]. It is important to be aware and understand the diversity of SCI as it makes it possible to deliver personalized therapy to each individual. At the same time, this understanding also gives an opportunity to obtain the perceptions of patients with respect to the effectiveness and practicality of an intervention [24]. Whereas conventional therapies offer some customization based on individual abilities, they are limited in being able to engage the neuromuscular unit in patients with severe paralysis or patients in the acute stages who are still in a state of flaccid spinal shock. FEST has the ability to overcome this and, via stimulation, we are able to engage the whole neural circuitry in functional movement's right from the time patients are medically stable to be engaged in rehabilitation. Whereas implanted devices require that patients "plateau" in function, surface FES devices can be used right from day one. Because of the versatile nature of these devices, stimulation programs can be modified and customized based on patient needs throughout the course of rehabilitation.

Among the FES devices researched in SCI, except for the Freehand [90] and NEC FES Mate systems, all other neuroprostheses for grasping are FES systems with surface stimulation electrodes. The NESS H200 [91], the Bionic glove [92], and the system developed by Rebersek and Vodovnik [93] are all surface stimulation devices that enable patients to perform hand opening and closing movements with assistance from FES, with the only difference being that the Bionic glove is almost exclusively tailored for the SCI population and enhances the tenodesis grasp in these patients. The Belgrade Grasping System [94] is capable of producing not only grasping function but also reaching by virtue of stimulation of the triceps muscle. The Compex Motion stimulator can be programmed to individual needs based on the level and extent of injury, and the patient goals to be achieved [95, 96]. MyndMove remains the state-of-the-art FDA approved transcutaneous stimulator for SCI with 8 stimulation channels and 13 pre-programmed stimulation protocols that allow for proximal and distal upper extremity retraining as well as combination protocols that allow more complex functional movements engaging the whole upper extremity.

2.5   OUR EXPERIENCES WITH SURFACE FEST TO RESTORE UPPER EXTREMITY FUNCTION AFTER STROKE AND SPINAL CORD INJURY

We have used FEST for over two decades in our laboratory to retrain function in individuals with stroke and SCI. We use a four-channel surface stimulator that is fully programmable. The parameters of stimulation used for upper extremity retraining are frequency of 40 Hz, pulse width of 250–400 μs and intensity between 5 and 40 mA. To date, approximately 150 individuals with SCI and 50 individuals with stroke have been treated using transcutaneous FEST, ranging from pilot clinical trials to randomized controlled trials. Our studies have included individuals throughout the continuum of care as well as with diverse impairment levels. For all of our clinical trials we have delivered 40 treatment sessions, with 5 sessions a week (1 session/day) or a minimum of 3 sessions per week (1 session/day). Our treatment strategy is to use FES during execution of functional day-to-day tasks using real objects and in adjunct with conventional strategies as seen appropriate by the treating therapist.

The general principles we follow during the FEST sessions are (as adopted from Kapadia et al. [97]):

1. identify the functions to be trained;

2. select the order of the tasks to be re-trained;

3. for each task identify the muscles to be stimulated;

4. first identify the optimal electrodes positioning;

5. apply self-adhesive electrodes over the motor points of the muscles identified;

6. identify and record the different stimulation thresholds;

7. explain to the patient what to expect when the FES is turned on;

8. turn on the stimulator and adjust the current intensities for all muscles to the levels determined previously;

9. instruct the patient that she/he has to make an active attempt to perform the intended movement and trigger stimulation once patient makes an attempt to super-impose their voluntary effort with the FES;

10. repeat this protocol 10–15 times;

11. rest time should be given when the patient asks for it and/or when muscle fatigue sets in; and

12. when the therapy is completed, turn off the stimulator, remove the electrodes, and inspect the skin underneath for any redness.

We have had very promising results treating individuals with both stroke and SCI [29, 35, 69, 70, 95, 96, 98] and our extensive research has led to the development of a commercial surface FES system that is now in use across several clinics and rehabilitation hospitals in North America.

2.6   LIMITATIONS AND CONTRAINDICATIONS FOR SURFACE FEST APPLICATION

The target muscles that are intended for FEST treatment have to be accessible for placement of the stimulation electrodes [99], i.e., surface FES can be applied with a greater level of specificity when the muscles to be stimulated are more superficial (e.g., *Flexor Carpi Ulnaris*, *Flexor Carpi Radialis*). For muscles that are deeper it might be more challenging to elicit isolated contraction without stimulating the overlying muscles (e.g., *Flexor Digitorum Profundus*). In some cases, such as the *Iliopsoas* muscle, it may not be possible to stimulate using surface electrodes owing to their location. Furthermore, the degree of lower motor neuron injury or nerve-root damage of the stimulated muscle should not be major. In a number of patients with SCI, there may be a variable amount of peripheral nerve damage [100] (motoneurons and nerve-roots) that restricts the application of FES. In addition, patients need to be able to follow the therapist's instructions and should not have any contraindications for use of FES. These include metal implants at the site of stimulation, pacemaker sensitive to electrical interference, open wound or rash at the site of electrode placement, and uncontrolled autonomic dysreflexia, among others.

> Disclaimer: The FES devices described here are not meant to be an exhaustive list of devices but rather a comprehensive description of the literature on surface FES devices used in rehabilitation of patients with stroke and SCI.

CHAPTER 3

Robotic-Assisted Rehabilitation

Over the past decades there has been a realization that the use of robots in rehabilitation seems not only a desirable innovation but perhaps also an inevitable one [101]. This is clearly demonstrated by an increase in volume of research in this field (Figure 3.1). As will be discussed in later chapters, robots have played a central role in the exploration of brain–computer interfacing technology as a rehabilitation tool. In this chapter, we review rehabilitation robots with historical importance as well as some that continue to be used today.

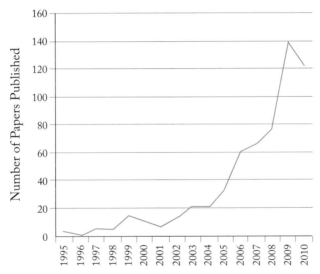

Figure 3.1: Growth in peer-reviewed articles published on rehabilitation robotics. Based on Ovid search on terms (rehabilitation or stroke or spinal cord injury or cerebral palsy) and (robotics or robot) performed on November 8, 2011 [101].

The interest in rehabilitation robotics has been motivated by the fact that they can reduce some of the physical effort required from the therapists [102]. The demand on the therapists stems from two important factors. First, the intensity of an intervention [103], which may refer to the frequency of sessions and the number of repetitions of a practiced movement during each session. Both of these are considered important factors that determine the efficacy of therapy. Second, specific interventions may require more than one therapist in each session. For example, practicing walking over a treadmill with body-weight support in spinal cord injured individuals may require two therapists ensuring the movement quality of each leg [104].

In addition, the use of a robot, often also makes it possible to quantify performance to monitor the progress of patients as an indicator of the efficacy of the therapy [102]. Hence the same device can "act" as an outcome assessment tool besides being a therapeutic device. The measurements may also assist in the process of documentation [105] and serve as an additional research tool [106].

Stroke is the best-studied condition from a rehabilitation robotics perspective, with the largest number of studies focusing on the upper limb. The robots used in stroke rehabilitation are categorized based on different conventions, such as the technical characteristics of each specific device, and the rehabilitation model that they support [107]. There is an overlap between these classifications. From a design perspective, robotic systems used for rehabilitation can be classified into two types [107, 108] according to how the movement is transferred from the device to the patient's extremities: end effector robots and exoskeletons.

3.1 END-EFFECTOR ROBOTS

End-effector robots make contact with a patient's limb at a single point, typically the most distal part, referred to as the *end-effector*. The interface between the robot and the limb can be realized with the use of a splint. Movements of the robot translate into changes in position for the entire limb [107, 109, 110].

3.1.1 END EFFECTOR UPPER EXTREMITY ROBOTS

MIT-MANUS

The MIT-MANUS is the oldest robotic system designed to assist, enhance, quantify, and document neurorehabilitation after stroke [105] (Figure 3.2). As the name suggests, it was developed at the Massachusetts Institute of Technology (MIT) to work in close physical contact with people [105]. The device allows unrestricted horizontal two-dimensional movements of the arm and shoulder [111] by incorporating a *selective compliance assembly robot arm* (SCARA). In addition, the MIT-MANUS has a module for assisting wrist movements. The system can move, guide, or perturb an arm in motion. This portable device can also measure position, velocity, and forces. When used, patients are asked to move their affected limb to control several video games on a screen. The MIT-MANUS is commercialized as the InMotion ARM® by BIONIK Laboratories Corp.

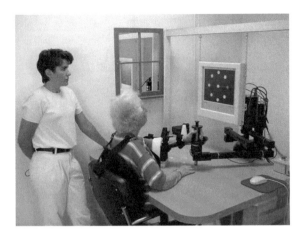

Figure 3.2: The MIT-Manus (image from [112]).

Mirror Image Movement Enabler (MIME)

The Mirror-Image Motion Enabler (MIME) is the result of a collaboration between the Department of Veteran Affairs Palo Alto Healthcare System and the School of Engineering at Stanford University [113]. The design of the robot also incorporates the input from individuals with disabilities who were involved early on in the development process. The robot allows practicing unrestricted active and passive movements in three dimensions, performed with one or both upper limbs. This is achieved by coupling a commercial robotic arm with six degrees of freedom with a splint that supports the person's hand and forearm (Figure 3.3). Individuals using this system are secured to a dedicated wheelchair equipped with a contoured backrest that restricts trunk movements while the robot is placed on a height adjustable table. The robot has four modes of operation: passive, patient-initiated active-assisted, active-constrained, and bimanual. The MIME makes it possible for individuals with stroke to practice bilateral movements by allowing the movements of a non-affected arm, recorded with a six degree-of-freedom digitizer, to control the movements of the impaired one [111].

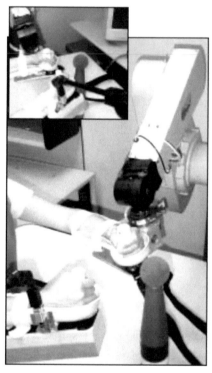

Figure 3.3: The Mirror-Image Motion Enabler (MIME) (image from [114]).

ARM Guide

The Assisted Rehabilitation and Measurement (ARM) Guide was developed during the 1990's by the Rehabilitation Institute of Chicago and University of California Irvine. The device was designed to guide arm reaching movements across a person's workspace (i.e., volume of space reachable by the hand), to measure range of motion and contact force generation, and apply controlled forces along linear reaching paths [115]. These features make it possible to measure arm multi-joint coordination and workspace deficits during reaching affected by a brain injury [116]. A person's forearm is attached to a splint which couples the individual's arm to a constraint that allows only linear movements. The orientation of the linear constraint can be modified in the vertical plane around the flexion/extension axis of the shoulder. As patients perform arm reaching movements along the linear constraint, the device measures forces and range of motion.

Bi-Manu-Track

The Bi-Manu-Track is a one-degree of freedom end-effector robot developed at the Neurorehabilitation Research Laboratory at Klinik Berlin/Charité University Hospital for the rehabilitation of individuals with severe upper limb impairments resulting from stroke [117]. The device makes it possible to practice bilateral forearm (pronation and supination) and wrist (flexion and extension) movements. To use it, patients sit in front of a table and place each of their forearms in a supportive guide. Each hand is placed around a handle (using a Velcro strap, if necessary) that allows rotational movements in either the horizontal or vertical axis to allow for movements of the elbow or wrist, respectively (Figure 3.4). The robot has several modes of operation including passive-passive, in which the robot moves both arms; active-passive, in which movements of the non-affected arm control the movements of the affected limb; and active-active, which requires that the impaired limb overcomes an isometric resistance to initiate movement).

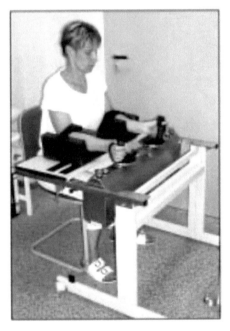

Figure 3.4: Bi-Manu-Track (image from [111]).

GENTLE/S

The GENTLE/S uses haptics and virtual reality technology. It was developed at the Human Robot Interface Laboratory, Department of Cybernetics, School of Systems Engineering, University of Reading, UK. The device components include a frame to support a person's arm and a three degree-of-freedom robot with haptic feedback (HapticMaster™) with a gimbal (also with three degrees of freedom) to support the forearm and wrist. In addition, the device incorporates a large computer monitor with screens [118]. A therapist can set the path of the movements to be practiced, the level of assistance delivered by the system, and the context for a virtual exercise environment. The device allows practicing pronation and supination of the elbow as well as flexion and extension of the wrist.

The NEuroREhabilitation RoBOT: NeReBOT

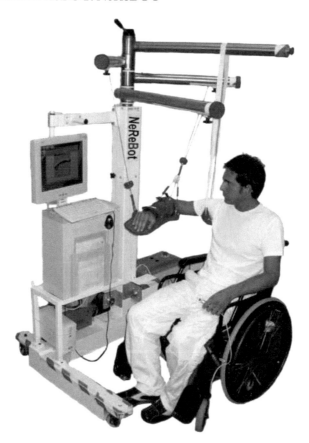

Figure 3.5: NeReBot (image from [120]).

The NEuroREhabilitation roBOT, commonly known as NeReBOT, is a wire-based robot with three degrees of freedom and provides visual and auditory feedback. The device was created at the Robotics Laboratory of the Department of Innovation in Mechanics and Management, University of Padua, Italy [119]. The robot is designed for upper limb rehabilitation, especially during the initial rehabilitation after stroke, characterized by muscle flaccidity [111, 119]. A rigid orthosis makes it possible to attach the robot's three wires to the patient's forearm [119] (Figure 3.5). The length of the wires is controlled independently by three electric motors. The shortening and lengthening of the wires make is possible to provide assistance to a limb moving in a wide working space [120]. The angular position and point of attachment for the wires of each of the NeReBot's links can be set by the operator. The robot allows patients to practice upper limb repetitive movements involving the shoulder and elbow (flexion and extension; adduction and abduction; pronation and supination; rotation). Patients can use the robot sitting on a wheelchair or from their bed, and the robot can be easily moved as it is supported by a wheeled frame [120]. The robot guides the patients by displaying a three-dimensional figure representing an arm with arrows indicating the intended movement to perform. In addition to facilitating movement, the NeReBot can also measure the patient's movements (speed and direction). Once the patient's arm is supported by the robot, the specific movements to be practiced can be defined by the operator by moving the person's arm manually and recording the angular positions of the robot's links.

3.1.2   END EFFECTOR LOWER EXTREMITY ROBOTS

Gait Trainer

The Gait Trainer is an end-effector type robot that simulates symmetric stance and swing phases of gait with a 60%–40% ratio, provides assistance to the participants, as required, and controls the movements of the center of mass [104] in the pelvis using ropes attached to a patient harness [121]. It allows patients undergoing rehabilitation to repeatedly practice gait-like movements with a decreased physical effort from the therapists providing care. The system consists of a double crank and rocker gear mechanism. A harness secures patients to an overhead frame, and their feet are attached to the device on two footplates (Figure 3.6). The Gait Trainer can provide complete or partial support. The Gait Trainer was also developed at the Klink Berlin/Charité University Hospital. Now in its second version, the device is sold by Reha Stim (Figure 3.7).

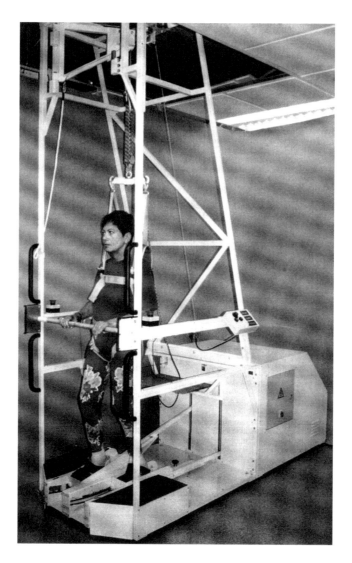

Figure 3.6: Early version of the GaitTrainer (image from [121]).

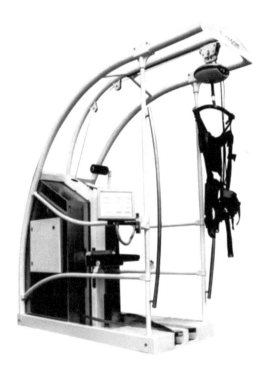

Figure 3.7: **Gait Trainer GT II** (https://reha-stim.com/gt-ii/).

HapticWalker

The HapticWalker is an end-effector robot developed to train arbitrary and freely programmable foot motions [122], making it possible to practice walking in even and uneven terrain. The system is designed as a haptic foot device, typically used to allow walking in virtual environments, in which each of the patients' feet is attached to footplates. The footplates are moved independently by two robotic modules that move the feet in the sagittal plane (Figure 3.8). The HapticWalker was developed at the Fraunhofer Institute IPK, Klinik Berlin/Charité University Hospital (Berlin).

Figure 3.8: HapticWalker (image from [117]).

3.2   EXOSKELETONS

Exoskeletons have a structure, which resembles the human upper limb, as robot joint axes match the upper limb joint axes [36]. These devices are designed to operate side by side with the human upper limb, and therefore can be attached to the upper limb at multiple locations [123]. Exoskeletons offer a larger range of motion (up to 7 degrees of freedom) compared to end-effector robots, with guaranteed optimal control of the arm and wrist movements [124]. These systems are suitable for the early stage of rehabilitation, as they do not require significant motor abilities.

3.2.1   UPPER EXTREMITY EXOSKELETONS

ARMin

The ARMin, created at the Sensory-Motor Systems Laboratory of the Swiss Federal Institute of Technology (ETH) Zurich, has four active and two passive degrees of freedom [125] (Figure 3.9). It allows elbow flexion and extension and shoulder movements in three dimensions. The ARMin is a semi-exoskeleton: an exoskeleton that facilitates the internal and external rotation of the upper arm and elbow joint. A wall-mounted end-effector-based component makes it possible to perform vertical and horizontal shoulder rotation. The device provides haptic, visual, and auditory feedback and provides only as much assistance while practicing movements creating a cooperative interaction with the patient. The ARMin has three training modes: mobilization, game training, and activities of daily living (ADL) training simulated in a virtual environment [126]. The ARMin technology is commercialized by Hocoma under the name ARMEO®Power.

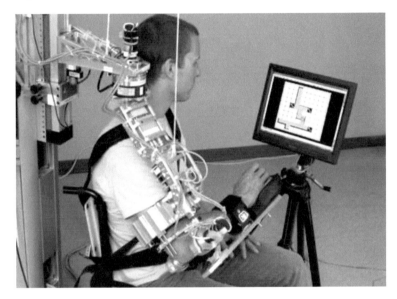

Figure 3.9: Early prototype of the ARMin (image from [125]).

Armeo®Spring

The Armeo®Spring is a passive spring-based exoskeleton that provides adjustable anti-gravity assistance/resistance to the arm [127] (Figure 3.10). The users' movements are recorded by the device and transmitted to a computer that provides real-time feedback in virtual reality-based video games that can be used during rehabilitation. The robot also allows for customization for each user by adjusting the range-of-motion required to operate the on-screen games. If suitable for the person undergoing rehabilitation, the device can also be used with a handgrip to incorporate finger flexion and extension practice (i.e., grasp and release during grip). The adjustable features of the Armeo®Spring, commercialized by Hocoma, make it suitable for a wide variety of upper limb impairments.

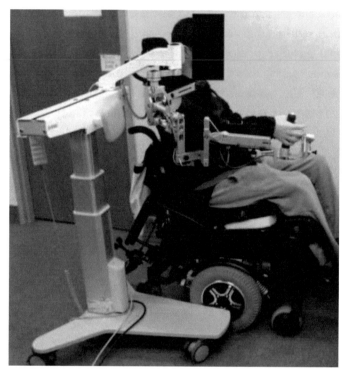

Figure 3.10: Armeo®Spring (image modified from [127]).

3.2.2   LOWER EXTREMITY EXOSKELETONS

Some of the earliest exoskeletons for rehabilitation applications were created in the 70's by Vuko-bratovic et al. to overcome the limitations of the existing systems at the time. The system consisted of an active orthosis actuated at the hip, knee, and ankles [128]. Since then, several other systems have been created with a notable increase in development and commercialization over the last two decades. As with their upper limb counterparts, lower limb exoskeletons assist the movement of a single joint or control multiple joints of a single limb.

Lokomat

The Lokomat was the first exoskeleton to support gait combined with a treadmill and body-weight support. It was designed to reduce the effort associated with manually assisted gait rehabilitation [111] for patients with spinal cord injury and stroke [106]. The device automates treadmill training, which requires that the patients' weight is partially unloaded with the use of a specialized harness, allowing practicing stepping and walking, typically with the assistance of at least two therapists (one assisting movements of each leg) (Figure 3.11). Developed at Balgrist University Hospital in the 1990s, the Lokomat has four degrees of freedom (left and right hip and knee joints) gait orthosis and a treadmill. The legs are attached to the device with the help of an upper leg brace and two lower leg braces. In addition to assisting both legs, as it is often necessary during SCI rehabilitation, it can also allow free-movement of one leg, suitable for individuals with hemiplegia resulting from stroke.

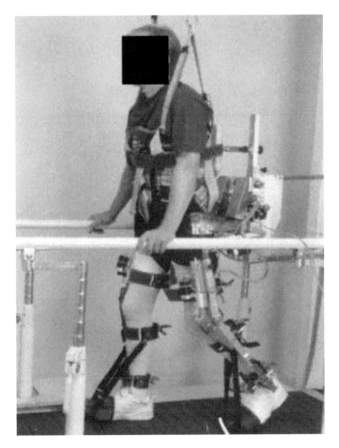

Figure 3.11: Early prototype of the Lokomat (image from [106]).

LOPES

The Lower Extremity Powered ExoSkeleton (LOPES) is a lower extremity powered robot with eight degrees of freedom (8-DOF) that combines a lightweight exoskeleton to facilitate leg movements and an end-effector that supports the pelvis [129] (Figure 3.12). The robot can provide active assistance during gait or allow unconstrained movements during treadmill training. Its design allows horizontal and vertical translational movements of the pelvis along with flexion, extension, abduction, and adduction. In addition to producing stepping movements with the exoskeleton, the LOPES allows the legs' abduction movements, which makes it possible to train lateral balance [102]. The system was developed at the Institute of Biomedical Technology of the University of Twente, Netherlands.

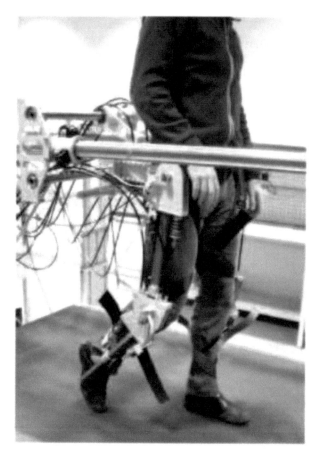

Figure 3.12: LOPES (image from [102]).

ReWalk

The ReWalk is an exoskeleton for individuals with SCI produced by ReWalk Robotics [130]. The device's orthosis is actuated at the hip and knee joints, and facilitates standing, turning, stair use, and overground walking. Users can activate the device using upper body movements. The ReWalk's controller and battery are contained in a backpack.

Ekso

The Ekso provides posture support and assistance during the swing phase of overground walking [131]. It can guide a limb in motion following predefined hip and knee kinematics providing constant assistance or as needed, according to the user's requirements. It also allows for undirected movements in the sagittal plane (i.e., users determine the moment in which the swing is initiated as well as the amplitude, velocity, and acceleration of the movement) for which it can provide assistance, resistance, or only gravity assistance. The Ekso, which has CE marking and is FDA approved, is manufactured by EKSO Bionics, U.S., with the EksoNR™ as its latest version, specifically designed for neurorehabilitation.

Hybrid Assistive Limb

The Hybrid Assistive Limb (HAL) is an exoskeleton to facilitate walking and can also assist during sit-to-stand and stand-to-sit transitions, as well as climbing and descending stairs [132]. The device is activated with electromyographic (EMG) signals obtained from flexor and extensor muscles of the knee (biceps femoris and medial vastus) and hip (rectus femoris and gluteus maximus). The exoskeleton is articulated at the ankle, knee and hip joints for each leg, each with one degree of freedom. The knee and ankle joints are actuated. The HAL has CE marking and is commercialized by CYBERDYNE Inc., located in Tsukuba, Japan.

3.3   CURRENT STATE OF EVIDENCE: ROBOTIC-ASSISTED THERAPY

3.3.1   ROBOTIC-ASSISTED REHABILITATION AFTER STROKE

There is an extensive body of literature, ranging from individual case reports to randomized-controlled trials, describing the use of robotic systems to assist rehabilitation of voluntary movement following stroke. The reports also have a significant amount of variability on the characteristics of the participants and the interventions (e.g., the type of robot used, duration and frequency of therapy, etc.). Consequently, it has been difficult to assess the efficacy of robot-assisted therapy in this constantly evolving field.

Upper Limb Rehabilitation after Stroke

Veerbeek et al. compared the effects of robot-assisted and non-robotic arm therapy across 44 randomized-controlled trials (with 1362 participants with stroke) [133]. Their analysis focused on motor control measures, muscle strength and tone, upper limb capacity, and activities of daily living. The results revealed a small but significant improvement in motor control and muscle strength, and the effects were specific to the joints targeted during rehabilitation. In terms of muscle tone, non-robotic interventions had better outcomes. There were no differences on upper limb capacity and activities of daily living.

In 2020, Chen et al. reviewed clinical trials that explored the use of robotic systems for rehabilitation of upper limb function in individuals with hemiplegia resulting from stroke [134]. Their analysis, which reviewed 35 trials, including 2,241 participants, found that the effects of robot-assisted therapy on motor impairment were slightly greater than those observed in treatments that only included a therapist, although the changes were not greater than the minimal clinically important difference (MCID). The effect was observed regardless of the trial design, the type of robot used, and whether the intervention focused on proximal, distal, or proximal and distal function.

In the same year, Mehrholz et al. provided a systematic overview of randomized-controlled trials using robot-assisted arm training and compared the relative effectiveness of the different devices and approaches used [135]. In total, their review included 55 randomized-controlled trials including 2,654 participants and compared 28 devices. The authors concluded that there was no statistical difference between robot-assisted therapy and best practice therapy for upper limb rehabilitation for improving participation in activities of daily living or hand-arm function.

A 2021 systematic review of randomized control trials exploring the use of rehabilitation robotics for unilateral rehabilitation after stroke suggested that interventions with end-effector type robots were superior to conventional rehabilitation for reducing impairment, especially when used to treat individuals with severe hemiplegia [136].

Lower Limb Rehabilitation after Stroke

Gait Rehabilitation

A 2017 review by Mehrholz et al. compared the proportion of participants walking independently after completing robot-assisted or best practice interventions for gait rehabilitation after stroke [137]. Their analysis included 62 studies consisting of randomized-controlled trials and randomized cross-over trials with a total of 2,440 participants. The studies were heterogeneous in the participants' characteristics (i.e., some were able to walk before the intervention), type of robot used, duration of the intervention, and use of other rehabilitation tools (i.e., some studies used functional electrical stimulation). In their findings, the combination of robot-assisted therapy and physiotherapy increased the likelihood of independent walking and improved walking velocity. Also, individuals in the first three months after stroke and those unable to walk are more likely to benefit from robot-assisted therapy.

Similarly, the 2019 Canadian Stroke Best Practice Recommendations, based on an extensive systematic review of the scientific literature, identify robot devices for gait training as a tool to be used in conjunction with conventional gait therapy for lower limb rehabilitation in individuals in the subacute and chronic rehabilitation stages who would otherwise not practice walking [23].

Balance Rehabilitation

A different report by Zheng et al. reviewed 31 reports, including 1,249 participants [138]. Their work collected evidence on the effects of robot-assisted therapy to improve balance function after stroke. Their analysis revealed that lower limb rehabilitation robots could improve balance over interventions that do not use this technology. Notably, the improvements were observed with most of the tested robots, and regardless of the intensity or duration of the intervention. However, the authors also encountered substantial differences in the reviewed studies suggesting a need for further research.

3.3.2   ROBOTIC-ASSISTED REHABILITATION AFTER SPINAL CORD INJURY

As mentioned earlier in this chapter, the number of studies exploring the use of robots for rehabilitation of SCI is small, when compared to the volume of work conducted in the stroke population. An important challenge in determining the efficacy of robot-assisted therapy after SCI is the fact that there are no existing standardized protocols [139]. It is likely that this also affects the clinical testing in the population of individuals who have had a stroke.

Upper Limb Rehabilitation after SCI

A scoping review in 2018 reported on the feasibility and outcomes of robotic-assisted training for upper limb function rehabilitation [139]. Their analysis included 12 articles including one randomized clinical trial, six case series, and five case studies, with a number of participants ranging between 1 and 17 for a total of 73. Their analysis revealed that individuals with mild-moderate impairments at the beginning of an intervention showed greatest improvements in body structure/function (smoothness of movement, pinch and grip strength, AIS upper extremity motor scores, and muscle strength). However, the authors also stated that the effectiveness of robot-assisted training was inconclusive and reiterated the need for more rigorous research. As with similar reviews, the authors identified the small number of articles, small sample sizes, the diversity of the devices, and their use, as well as their outcome measures as limiting factors. For example, the authors of a different review in the previous year comparing the efficacy of robot-assisted therapy with other interventions reported that they were unable to perform an analysis on interventions focused on the upper limb due to a lack of studies [140].

Lower Limb Rehabilitation after SCI

Cheung et al. reviewed 11 randomized-controlled trials and quasi-randomized-controlled trials comparing robot-assisted therapy with other interventions for rehabilitation of lower limb voluntary function [140]. The effects of robotic therapy, assessed through the 443 participants included in the authors' synthesis, appeared to be greater for improving walking independence and endurance with incomplete SCI in both acute and chronic stages of rehabilitation.

In 2020, Fang et al. reviewed the effects of robot-assisted gait training on pain and spasticity, which commonly afflict individuals with SCI [141]. In total, the authors included 18 studies in their analysis with 301 participants. No improvements were observed on pain. In addition, robot-assisted therapy improved spasticity in the 11 studies (non-randomized-controlled trials) as did lower extremity motor score and walking ability.

3.4   ROBOTIC AND FES HYBRID SYSTEMS

The combination of FES and robotic devices into a single system for neurorehabilitation is motivated by an interest to overcome the limitations associated with each technology [142]. Delivering power to a robotic system is a significant challenge, and some of them are also non-portable. Important problems associated with FES revolve around muscle fatigue and poor trajectory control of the resulting movement. Combining FES and robotic systems provides an opportunity to reduce energy requirements by, for example, using FES as a power source for the movement and a mechanical orthosis (i.e., exoskeleton) to guide the movement and reduce muscle fatigue by providing mechanical support (e.g., during standing). Hence, hybrid FES and robot systems may increase intensity during training with more time spent practicing the activity and ensuring activation of the correct muscle groups involved in the practiced task.

3.4.1   ROBOTIC—FES HYBRID SYSTEMS FOR WALKING

Two main approaches have been explored throughout the development of hybrid systems combining FES and robotic technologies to facilitate walking. These include exoskeletons controlled by breaking mechanisms in specific joints and systems that have actively actuated joints [142].

3.4.2   HYBRID SYSTEMS WITH JOINT-BREAKING

Some of the approaches to implement hybrid exoskeletons controlled by joint breaks have included combining a passive hip-knee-ankle-foot orthosis (HKAFO) with a 16-channel implanted FES system [143, 144]. The orthosis incorporates a variable hip constraint mechanism that uses spring clutches to lock and release the hip and knee joints. For example, during standing, the knee joint is locked while allowed to move during the swing phase.

A different approach combines a four-channel FES system with an orthosis with eight degrees of freedom equipped with magnetic breaks on the hip and knee joints [145]. Electrical stimulation is delivered to the quadriceps to produce knee extension and the perineal nerve to elicit the withdrawal reflex. The breaks control the position and velocity of the resulting movements. The device reduces muscle fatigue by controlling the intensity of the FES stimulation using torque and trajectory information. Similar approaches have been implemented in which the stored energy acts on the knee and hip [146] and incorporating pneumatic accumulators to transfer energy to the hip [147, 148].

3.4.3   HYBRID SYSTEMS WITH ACTIVE JOINT CONTROL

The main drawback of the hybrid exoskeletons with joint braking is their inability to provide complete joint control since joint brakes are not capable of delivering the necessary torque; movement quality is low in terms of joint trajectory and velocity. Therefore, in contrast to joint brake hybrid exoskeletons, active actuator hybrid exoskeletons can control the power delivered at the joint and allow effective closed-loop control of joint movement.

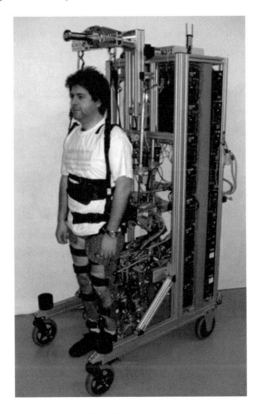

Figure 3.13: The WalkTrainer (image from [149]).

One of the first exoskeletons with actively actuated joints was the Hybrid Assistive System (HAS) [86], also mentioned in the previous chapter. The HAS provided active assistance using DC motors, making it possible to flex and extend the knee, and could also lock it. A control system coordinated the actuated orthosis with an FES channel that assisted with balance, hip flexion and knee extension and the withdrawal reflex.

The WalkTrainer combines an FES system with an exoskeleton and a moving frame with motorized wheels that provides body-weight support while walking [149] (Figure 3.13). The exoskeleton assists hip, knee, ankle, and pelvis movements. The system estimates the torques on each joint to adjust the intensity of the stimulation.

3.4.4   ROBOTIC—FES HYBRID SYSTEMS FOR UPPER LIMB FUNCTION

Similar to hybrid systems for lower limb function, the systems designed for the upper limb often consist of an FES system to produce movement through the activation of the users' own musculature and an exoskeleton that can provide stabilization of the moving limb, anti-gravity support, or active movement.

The MUltimodal Neuroprosthesis for Daily Upper limb Support (MUNDUS) is a hybrid device for restoring reaching and grasping that accepts a wide variety of signals for control and feedback [150]. Among the modalities that can be used to trigger the MUNDUS are a push-button, electromyographic signals, eye-tracking, and EMG signals. Motion signals, obtained from encoders included in the exoskeleton or an instrumented glove, enrich the environmental information provided by a depth camera and RFID-tagged objects. The exoskeleton has one degree of freedom at the elbow and three degrees of freedom around the shoulder, with only the ones associated with shoulder flexion/extension and adduction and abduction being mobile (Figure 3.14). The system uses two stimulators to facilitate movements of the arm, forearm, and hand. Finally, the device can also be integrated with a robotic hand orthosis.

Ambrosini et al. developed a hybrid system for upper limb rehabilitation, the RETRAINER-ARM, that combines an EMG-triggered FES system with a passive exoskeleton which provides anti-gravity assistance [151] (Figure 3.15). The exoskeleton, which can be mounted on a wheelchair, has three degrees of freedom that allow for shoulder rotation and elevation as well as elbow flexion/extension. Each of the joints is equipped with an electromagnetic brake and can also provide angle measurements. A fourth uninstrumented joint compensates for trunk movements. The device includes a four-channel electrical stimulator that is used to produce arm movements. Two of the channels can be used to measure EMG activity and trigger the stimulation as well. Of note, the hybrid system is also designed to operate with interactive physical objects that provide proximity information, similar to the MUNDUS [150].

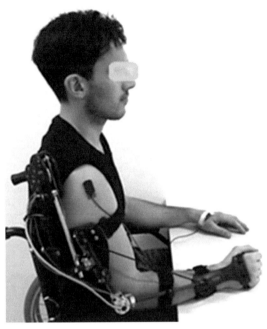

Figure 3.14: Exoskeleton used by the MUNDUS hybrid system (image modified from [150]).

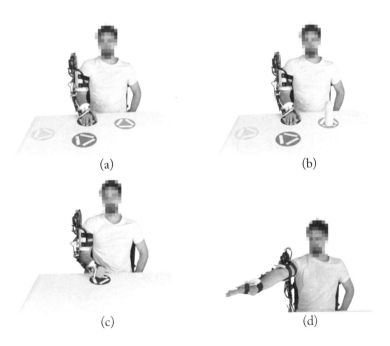

(a)   (b)

(c)   (d)

Figure 3.15: The RETRAINER-ARM hybrid system can facilitate arm movements (image modified from [151]).

The device developed by Scott et al. combines a robotic system that provides active and passive flexion and extension of the hand and passive range of motion to the thumb and fingers [152] (Figure 3.16). The system also facilitates the movements using FES.

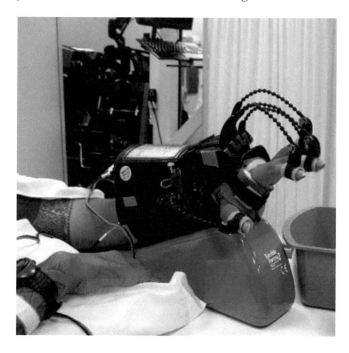

Figure 3.16: Hybrid system for rehabilitation of hand function developed by Scott et al. (image modified from [152]).

3.5   CLOSING STATEMENT

In conclusion, evidence in literature points to the fact that although rehabilitation robots are in a dynamic phase of development and have achieved impressive advances technologically its adoption in routine clinical care is yet to be realized. Continued improvements in the technology, more data on efficacy, and efforts to develop truly labor-saving devices have the potential to bring rehabilitation robotics fully into the clinical realm. Ultimately, the ability of robotic devices to reduce the personnel costs associated with optimal rehabilitation programs will be the driving factor that results in their widespread adoption [153].

All of the robotic rehabilitation devices discussed in this chapter are rigid robots, and these, by virtue of their structure, are limited in their ability to adapt to the multidimensional fluid movements of human limbs. To address this issue, a promising upcoming field is soft robotics. Although a detailed literature review of this new soft robotics technology is outside the scope of this book, it becomes imperative that we introduce the reader to this rapidly growing research field. The mechanical properties of the materials used in the construction of soft robots may allow the creation of devices that are not only safer but also provide a better anatomical and functional fit. This could offer an unprecedented level of versatility to facilitate movements while practicing complex tasks. [108].

Overall, robotics is a promising field of research that has the potential to revolutionize the way we provide therapy and, just like many other research endeavours, requires brainstorming from multidisciplinary teams to make it viable in clinical settings.

CHAPTER 4

Brain–Computer Interfaces

Although the notion of controlling the environment using the activity of the brain alone may have been around since the first reports describing electroencephalographic (EEG) recordings [154], one of the earliest descriptions of a brain–computer interface (BCI) was produced in 1977 by Jaques Vidal [155] from the University of California Los Angeles (UCLA). In this early example of BCI technology, changes in EEG activity, elicited by looking at four flashing lights—each representing a different direction and placed on a diamond-shaped panel—allowed the user to control the movement of a virtual cursor to navigate a maze. In addition to its relevance as one of the first reports formally describing a BCI, this work was also important as it introduced fundamental elements and problems of this technology. This chapter provides a general overview of brain–computer interface technology and its applications.

4.1 DEFINITION

A brain–computer interface, or BCI, can translate brain signals into control commands for electronic devices. Jaques Vidal used this term in the early 1970s [156] for the first time. Other names for this technology include brain-machine interface (BMI) and direct brain interface, although there are often procedural differences (related to the techniques used to record the activity of the brain) that have distinguished these terms. Today, the term *BCI* often refers to applications in which brain signals control computers while *BMI* describes any other application.

During the first international meeting on brain–computer interfacing, held in June of 1999, a group representing 22 international research groups created the following definition: "A brain–computer interface is a communication system that does not depend on the brain's normal output pathways of peripheral nerves and muscles." [154]. Today, 20 years later, this definition continues to be appropriate. However, it has become clear that the specific aspects of the communication supported by the technology and the brain's output are shaping BCI technology's evolution.

4.2 COMPONENTS OF A BCI

A BCI has an input (i.e., a brain signal), an output (e.g., a command for an electronic device), and an intermediate stage that converts the input into the output (Figure 4.1). This intermediate stage often contains two portions, the first to extract information from raw brain activity and the second one to translate the extracted information into a suitable representation for its use as a control signal. Each of these components is an active area of research in the BCI field, along with the applications of this technology. During use of a BCI, the resulting action provides immediate feedback. The user can use this feedback to learn how to modify his or her brain activity.

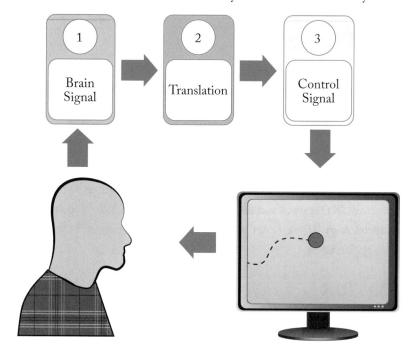

Figure 4.1: A brain–computer interface has an input (i.e., a signal representing the activity of the brain), an output (i.e., a command for an electronic device), and a processing stage that extracts relevant information from the brain signal(s) and translates it into the control signal. Feedback (i.e., the resulting action) is immediately available which helps the user learn to modify his or her brain signal voluntarily.

4.2.1   SIGNALS FOR IMPLEMENTING BCI SYSTEMS

Virtually any technique that exists to record the brain's activity can and has been used to implement BCI technology. The different recording methods are categorized broadly according to whether or not the method is invasive. Thus, sometimes this is also used to classify BCI systems (i.e., invasive BCI vs. non-invasive BCI).

Invasive Recordings

As the name suggests, invasive methods to acquire the activity of the brain require the placement of electrodes penetrating brain tissue or directly in contact with the surface of the brain. These techniques have a spatial resolution that is superior to that of non-invasive recording methods. They are also generally considered to have a broader frequency content. Also, the time required to set up the system for operation may decrease significantly. However, a few of the disadvantages of invasive systems lie on the fact that they require surgery for their implantation, which increases cost and the risk of infection. The two main invasive recording techniques used for BCI/BMI implementation use microelectrodes and subdural electrodes.

Intracortical Microrecordings

One crucial technique to record brain activity consists of the use of electrodes on the micron scale. These microelectrodes make it possible to record the activity of neurons both individually and as part of a population. The electrodes are placed in the cerebral cortex over specifically targeted areas displaying activity correlated with specific behavior such as arm movements.

The use of intracortical electrodes in humans, placed over the primary motor cortex, has made it possible to operate a computer cursor in simulated applications to send email and interact with the environment (i.e., electronic aid for daily living) [157]. Intracortical recordings have also allowed predicting imagined movements performed with a single joint (including shoulder, elbow, forearm pronation and supination, wrist, and hand opening and closing) [158], control of a virtual arm [159] as well as an anthropomorphic robotic arm with seven degrees of freedom [160], and more recently, control of a virtual arm and real arm using electrical stimulation [161].

Subdural Recordings

Neurologists specialized in movement disorders use subdural recordings to observe patients with intractable epilepsy. The procedure involves placing an array of electrodes on the surface of the brain underneath the dura mater. The clinical use of subdural electrodes makes it possible to determine the site of origin of a seizure. A medical team can later use this information to decide if surgical removal of the epileptic foci is possible.

Subdural electrodes have been used to create BCI/BMI systems. Neurosurgeons often describe the surgery for implanting the electrodes as minimally invasive, and Electrocorticography (ECoG) signals are often regarded as having a broader spectral content. Importantly, subdural recordings contain information on the kinetic and kinematic attributes of movement including direction [162–166], speed, velocity [166], and range of motion [162] of a hand in motion, as well as flexion of individual fingers [167].

Subdural recordings have made it possible to create systems that can reconstruct kinematic or kinetic parameters of voluntary movement or identify performed or intended movements from ECoG signals alone. ECoG-based systems have also been used to control movements in two dimensions [168–170] and electrical stimulation devices for restoring upper limb movements [171].

Epidural Recordings

In addition to intracortical and subdural recordings, there have been reports describing BCI systems using epidural recordings. In this case, surgeons place the electrodes on the dura mater instead of on (or inside) the brain. One important rationale for using epidural recordings is that the surgery is less invasive, resulting in reduced trauma to the person receiving the electrodes. Epidural record-ings have made it possible to identify the intention to perform multiple hand movements [172].

4.2.2   NON-INVASIVE RECORDINGS

Non-invasive recording techniques do not require a surgical implantation for their use. The signals represent measures of activity using blood flow, magnetic, or electric measures [173] and are de-scribed briefly next.

Functional Magnetic Resonance Imaging (fMRI)

Functional MRI (fMRI) can demonstrate local time-varying metabolic changes in the brain as re-flected in increased energy consumption (in the form of adenosine triphosphate—ATP) during in-creased neural activity [174]. More specifically, increased brain activity can result in a hemodynamic response that involves the increase and decrease of oxygenated and deoxygenated hemoglobin as a result of neural activity. Given that fully oxygenated hemoglobin is magnetically indistinguishable from brain tissue while deoxygenated hemoglobin is highly paramagnetic, magnetic resonance imaging (MRI) techniques make it possible to visualize these metabolic changes as they take place.

Functional Near Infrared Spectroscopy (fNIRS)

Functional Near Infrared Spectroscopy (fNIRS) is another imaging modality for brain hemody-namics [175]. In contrast to fMRI, which uses changes in magnetic properties of tissue, fNIRS takes advantage of the fluctuations in the tissues' optical properties, primarily blood absorbance, that reflect neural activity. To do this, fNIRS systems use light outside of the visible range (wavelengths between 750 nm and 1,200 nm), which is delivered using optodes placed on a person's head. The signals obtained with this imaging modality are indicative of the concentrations of oxygenated and deoxygenated hemoglobin [176]. This makes it possible to measure relative increases in neural ac-tivity which increases oxygen demands and consequently blood flow. In the context of BCI research, systems based in NIRS have been used to discriminate between mental arithmetic and mental singing [177], among other applications.

Magnetoencephalography (MEG)

Magnetoencephalography is a brain imaging modality that senses the magnetic induction produced by the electrochemical currents that transfer information between neurons. The magnetic induction induces a measurable proportional electric current in a pickup coil. The coil magnetometers in a MEG system are arranged so that they cover the entire head, forming 300 channels, and do not come in contact with the person. Like an fMRI system, MEG devices are not portable and can be expensive to operate. MEG has made it possible, among other things, to identify the intention to perform arm [178] and hand [178, 179] movements.

Electroencephalography (EEG)

The vast majority of BCI systems use the electrical activity of the brain. The most common technique to record this activity is EEG, first reported in humans in 1929 [180]. The non-invasive nature, ease of recording, and relatively low cost, make this long-standing approach to recording brain activity ideal for neurorehabilitation. EEG signals have an amplitude of less than 1 mV and a frequency band that ranges from 0 Hz (DC) to 100 Hz [181]. Scientists believe that these signals reflect the combined activity of hundreds or thousands of pyramidal neurons, organized in columns from the outer to the inner layers of the cerebral cortex; this parallel organization contributes to the summation of the neuronal activity [181]. Acquisition of EEG signals requires the use of electrodes placed on a person's scalp. The electrodes are placed individually or as part of a garment (i.e., an EEG cap). Electrodes are typically positioned following the 10–20 electrode placement system which allows placing the electrodes over different anatomical areas of the brain (i.e., different cortices). It receives its name from the fact that it uses percentual fractions (10% and 20%) of the distances between four anatomical landmarks of a person's head. The extended 10-20 electrode placement system, which is based on 10% anatomical distance fractions resulting in a higher number of electrodes, is also used commonly for BCI implementation. Labeling of each electrode follows an alphanumerical nomenclature consisting of one letter and one number. The first of these characters refers to the cerebral lobe underneath the electrode and can adopt the values Fp, F, T, P, C, and O, referring to pre-frontal, frontal, temporal, parietal, central, and occipital. Please note that there is no central lobe, and the letter C only identifies the central portion of the cortex. A second character is a number which identifies the hemisphere: odd (1, 3, 5, and 7) and even (2, 4, 6, and 8) numbers refer to the left and right hemispheres, respectively, and a lowercase letter "z" represents zero (i.e., the midline defining the sagittal plane).

4.3   OTHER CLASSIFICATIONS OF BCI SYSTEMS

4.3.1   ENDOGENOUS VS. EXOGENOUS

Another approach to classify BCI systems is based on whether or not an external stimulus is required to produce a change in brain activity that can be identified. *Exogenous* systems require the application of a sensory stimulus that can produce a measurable change in brain activity, making it possible to implement the BCI while *endogenous* BCI systems use brain signals that are generated internally [182].

Exogenous systems may require, for example, presenting the user with a flashing light. This flashing light will result in a change in brain activity, steady-state visual evoked potentials (SSVEP), which are reflected by oscillations recorded in occipital areas that match the flashing frequency. This was indeed the strategy used by Jaques Vidal [155], mentioned at the beginning of this chapter. In addition to SSVEP, which continues to be used today, another potential used frequently for implementing exogenous BCI systems is the P300 potential, produced by an unexpected presentation of a relevant (i.e., meaningful) stimulus. The P300 potential is described in greater detail in Section 4.6.1.

In contrast, endogenous systems use activity that reflects an internal mental state of the user. An endogenous BCI may require, for example, performing mental arithmetic or mental singing [177]. Importantly, motor imagery is also used extensively to create endogenous BCIs. These tasks induce changes in the activity of the brain that can be identified and translated into a control command.

4.3.2   SYNCHRONOUS VS. ASYNCHRONOUS BCI SYSTEMS

Another way to classify BCI systems revolves around whether or not the BCI determines when the user can issue a command. *Synchronous* systems monitor the activity of the brain only during specific time windows defined by the BCI itself. The user can only issue a command (i.e., activate the BCI) during these periods. Consequently, the technology controls the timing and speed of interaction between the user and the environment. Most BCIs are implemented using this synchronous approach.

In contrast, *asynchronous* systems (also referred to as *self-paced* BCIs) allow users to issue a command any time, without any constrains. To do this, the system monitors the activity of the brain constantly with the goal of identifying signatures of intended control. Increasing the difficulty of implementing these systems is the fact that the periods of no-control [183] are usually undefined; the user may be engaged in multiple activities, not accounted for specifically when designing the BCI. Researchers often consider asynchronous BCIs better suited for real-world applications, outside of a research environment.

4.4   CONVERTING BRAIN ACTIVITY INTO A CONTROL SIGNAL

The intermediate component of a BCI, as stated earlier in this chapter, is dedicated to converting the activity of the brain into a control command. This makes use of extensive signal processing and automatic classification tools that are beyond the scope of this book. However, the two fundamental tasks performed in this stage include extracting information from the acquired brain activity (i.e., feature extraction), indicative of the user's intention, and converting the extracted information into a representation suitable for its use as control signal. Extraction of information from acquired brain activity typically includes a preprocessing stage to increase the quality of the recordings. This signal enhancement can be achieved using signal processing techniques including, among multiple others, independent component analysis (ICA) and spatial filtering such as the surface Laplacian, which accentuates activity surrounding a specific sensor while reducing common activity present in all of the remaining sensors. Other often used pre-processing methods include common average referencing (CAR), common spatial patterns (CSP), and principal component analysis (PCA).

Extracting features that make it possible to identify specific brain events/states that convey information about the user's intention is dependent on the associated neurological mechanisms. Researchers have tried and created multiple approaches for extracting features throughout the history of BCI development. Some of the feature extraction techniques used for BCI implementation include spectral parameters, parametric modelling, cross-correlation, amplitude measures, and many others.

Translation of the extracted features can also take multiple forms. In some cases, the feature itself (e.g., power in a specific EEG frequency band) can be simply mapped to an equivalent control signal. A different approach is to implement a detector based also on the power level of an EEG signal or the shape of a particular potential. And finally, the features can also be classified. Common classification approaches used include linear classifiers, such as linear discriminant analysis (LDA) and support vector machine (SVM), neural networks, Bayesian classifiers, and distance-based (nearest-neighbor) classifiers.

4.5   MOTOR-RELATED EEG FEATURES FOR BCI IMPLEMENTATION

Voluntary movement produces changes in the activity of the brain as early as two seconds before the movement takes place. These features have been used extensively to create BCI technology and play a central role in its integration into rehabilitation of voluntary movement. They are described next.

4.5.1   EVENT-RELATED DESYNCHRONIZATION (ERD)

One of the features used for implementing BCIs is a decrease in power commonly referred to as event-related desynchronization or ERD and observable during the execution of voluntary movement. This phenomenon can be recorded using EEG, ECoG, and MEG techniques over sensorimotor areas, and more prominently on the side contralateral to the moving limb. The widespread use of ERD for implementing BCIs is due to its presence prior to the onset of movement, while imagining movement, and when a movement is attempted; no overt movement is required. Scientists have considered ERD as a decrease in synchrony in neighboring neuronal populations and have described this desynchronization as an indicator of cortical activity associated with the production of movement and processing of sensory information [184].

Calculation of ERD requires spectral analysis, and it is present in different frequency bands, often within the mu (8–12 Hz) and beta (13–30 Hz) ranges, for each individual. The sites displaying ERD are also user-specific. A typical approach to measuring ERD starts with acquisition of multiple trials in which users imagine or attempt a movement. The movements are usually performed following a sequence of "READY", "GO", "REST" cues, with a pause of several seconds between repetitions. After the data is collected, the trials are aligned to a specific event, such as a "GO" signal, and trimmed, ensuring that the resulting segments contain activity before and after the event used for alignment. In its simplest form, the resulting trials are processed using the following steps:

1. Apply a filter to extract the frequency of interest (e.g., a sub-band within the alpha frequency range).

2. Estimate the power of the signal by squaring each sample.

3. Average the resulting power estimates across all trials.

4. Apply a smoothing filter.

Finally, ERD is defined as the percentual change with respect to a baseline period. This relative expression of ERD is calculated according to:

$$ERD = ((A - R)/R) \times 100$$

where

A = period after the event used for alignment and R = power in the baseline period. The process is illustrated in Figure 4.2.

It is also possible to create ERD maps, which allow inspecting the temporo-spectral and spatial characteristics of brain activity [185]. The process described above is repeated for multiple frequency bands and electrodes. The results are then presented as a three-dimensional topographical map to visualize which EEG electrodes and frequency bands display ERD and are suitable for implementing a BCI. It is also possible to display only values that show a significant statistical difference from the baseline period, which aids in the interpretation of the plots. An example of this process is shown in Figure 4.3.

It is important to mention event-related synchronization, which is an increase in power in the oscillatory activity of the brain. As with the desynchronization, ERS is also associated with voluntary motor function, and can be observed with multiple techniques for measuring brain activity. For example, it is common to observe ERS in the beta frequency band after completing a movement or imagining it.

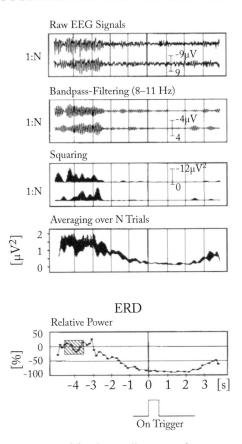

Figure 4.2: Example of ERD calculation. The figure illustrates the steps to calculate ERD in the 8–12 Hz frequency range. Image modified from [184].

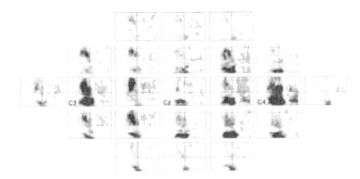

Figure 4.3: Example of an ERD map. The figure illustrates the ERD maps calculated for 23 EEG channels. with only significant ERD values displayed. Image modified from [185].

4.5.2   MOTOR-RELATED CORTICAL POTENTIAL (MRCP)

A second EEG feature for implementing BCI that uses activity associated with voluntary movement is the motor related cortical potential (MRCP). Other names that may be used include contingent negative variation (CNV) and Bereitschaft potential (BP) although, MRCP often encompasses all of the potentials related to movement [186] (including CNV and BP).

The MRCP is a slow cortical potential consisting of a slow negative shift that can be observed in EEG activity at frequencies below 5 Hz [187]. The potential reflects activation from the premotor cortex and the supplementary motor area during cue-based and self-initiated movement [188]. The maximum negativity of the MRCP is associated with the onset of real or imagined movements [187].

Relevant to the implementation of BCI systems, MRCP can be used to detect the intention to move, or imagine movements, several seconds before onset [189]. Also important is the fact that operation of MRCP-based BCIs does not require training [188], as is the case with many systems that use ERD.

Like other potentials, the amplitude of the MRCP is small, compared to ongoing EEG activity. Because of this, calculation of MRCP requires averaging over multiple trials (e.g., 40–50 [190]) aligned with respect to an experimental cue or the onset of movement, if available. Spatial filtering may also be used by subtracting the average activity of the surrounding electrodes, which enhances the MRCP [191, 192].

4.6   APPLICATIONS OF BCI TECHNOLOGIES

4.6.1   AUGMENTATIVE COMMUNICATION

The development of BCI technology has been strongly motivated by the desire to assist individuals with limited ability to move voluntarily. Specifically, early BCI reports often describe the potential of this technology to assist locked-in individuals. Patients with this condition lose the ability to perform voluntary movements, often completely, while retaining all other cognitive abilities. This situation makes not only speaking difficult or impossible but also precludes the use of other alternative and augmentative communication devices that may rely on other voluntary movements (e.g., facial movements, eye gaze, blinking). Consequently, one of the initial intended applications and strong motivation for the development of BCI systems was facilitating communication in this population.

An important example of these early systems is the *Mental Prosthesis* [193]. This BCI made it possible for users to type using the P300 potential, a positive response, observable in EEG activity, that takes place 300 ms after a stimulus. The stimulus used is an event that is meaningful and rare for the person using the system. In the case of the Mental Prosthesis, a table displayed the alphabet and a few punctuation marks. Each row and column would flash randomly. To select a character, the user counted the number of times that the desired character flashed (making the flashing events relevant). Given that a P300 response was produced only by the desired character, the BCI could determine which character (i.e., the element on the matrix) was the intended one.

Research and development of P300-based BCI's continue today with applications beyond text production. More recent work has included robotic, orthotic, and environmental control, and it has also seen the integration of BCIs with augmented reality systems [194, 195] as well as tactile stimulation [196].

4.6.2   COMPUTER ACCESS

Another important early application was the control of personal computers using BCI technology. In particular, one of the applications that received attention was the control of a pointer (mouse), arguably because the majority of computers used by the general public use a graphical user interface. As such, there are multiple examples of BCI implementations to control a pointer moving in one- and two-dimensions using EEG [197–200], ECoG [165, 168, 170, 201], and intracortical recordings [157, 203].

4.6.3   ADDITIONAL DEMONSTRATION APPLICATIONS

In addition to assistive technology applications, there has also been a series of feasibility demonstration projects showcasing innovative BCI technology applications. Some examples include control of a quadcopter drone [204], control of a cursor in three-dimensional virtual space [205, 206], control of a humanoid robot [207], control of a robotic arm using intracortical [160, 209] and EEG signals [210], and control of a transcranial magnetic stimulator to produce a motor response in a different person (i.e., a brain-to-brain interface) [211].

4.6.4   FACILITATION OF MOVEMENT AFTER PARALYSIS

Eventually, applications emerged in which BCI systems were used to control technologies specifically designed to restore movement. The development of these applications was perhaps not surprising given that the motivation of BCI research was to assist individuals with disabilities.

BCI-Triggered Actuated Orthoses

Some of the first examples of BCI applications for the restoration of movement used actuated orthoses. In one of the earliest examples of control of a device to facilitate movement after paralysis, Pfurtscheller et al. reported using a hand orthosis by an individual with chronic tetraplegia (C4/C5 level sustained approximately 12 years before the report date) [212]. The participant imagined moving his feet to close the left-hand orthosis while opening the device required imagining moving his opposite (right) hand. Imagined foot movements elicited an EEG power increase in the 16–18

Hz range (beta) over mid-central (Cz) regions easily distinguishable from imagined movements of the right hand using LDA classification. The onset of this beta activity became faster (i.e., appeared earlier) as the participant used the device, indicating potential neuroplastic changes associated with the regular practice. The user could use the system to perform functional tasks (e.g., eat an apple).

BCI-Control of Prosthetic Limbs

Another important example was provided by Hochberg and his colleagues, who reported using intracortical activity to control the closing and opening of a prosthetic hand [157]. The study participant had tetraplegia (C4) resulting from a traumatic SCI sustained three years before the research took place. The research team implanted an array containing 96 microelectrodes placed over the motor cortex. The firing frequency of individuals cells in the neuronal ensemble was transformed into a continuous two-dimensional control signal using linear filters. The system enabled the participant to control a computer cursor as well as a robotic arm. The same system also allowed control of the prosthetic hand with a few modifications.

Another system using intracortical recordings controlled a prosthetic arm with multiple degrees of freedom [160]. Activity obtained using a 96-channel microelectrode array was used to control translation and orientation in space as well as closing the manipulator to perform grasping. A person diagnosed with a spinocerebellar degeneration used the system effectively to manipulate several small objects (a ball, a rock, and blocks and tubes of different sizes). Today, the use of BCI technology to restore movement continues to be an important area of research. It has now expanded to the control of exoskeletons designed to facilitate walking in individuals with SCI.

Virtual Reality

Although not strictly for facilitating movement, virtual reality has provided the necessary conditions to develop BCI technology for mobility in a safe environment. In other cases, virtual reality has also provided an initial testbed before the full implementation of complex applications for movement restoration. The results of the work conducted in virtual spaces have often led to the eventual creation of physical equivalents of BCI technology.

In 2007, Leeb and colleagues reported the control of movement inside a virtual environment [183]. The study included one participant living for several years with an SCI at the cervical level (C4 incomplete) resulting in tetraplegia. Following a training regime lasting four months, the participant was able to produce, on demand, bursts of activity in the beta frequency bands over medial central areas of his head by imagining foot movements. This allowed the experimenters to create a switch-like BCI suitable for asynchronous applications. During the experiments, the participant sat inside a cave virtual reality system, consisting of a multi-projection room that provided an immersive reality experience. That participant was able to move to 15 avatars, making sure to stop briefly for each one of them.

A group of researchers at the University of California, Irvine, described a system to control walking in a virtual reality environment [213]. Using 63-channel EEG recordings, 9 participants, including one individual who had lived with paraplegia due to SCI (T8) for 11 years, were able to walk and stop between 10 virtual avatars standing in a straight line. Notably, the system required only 10 min to allow the participants, and the BCI, to become operational. The results of that work suggested the feasibility of BCI systems for restoring lower-limb function. The research team later reported the same system used successfully by five additional individuals with chronic SCI, including four living with paraplegia (all at the T1 level) and one person with tetraplegia (C5, syringomyelia) [214].

Researchers at Case Western Reserve University demonstrated the real-time control of a virtual arm with two degrees of freedom. The participant of the study was a 56-year-old woman with tetraplegia following a brainstem stroke. She had no functional movement of her arms and legs and her somatosensory function was intact. She was also unable to speak. She was implanted with an intracortical array in her left primary cortex (precentral gyrus) over the representation of the arm. At the beginning of each experimental session, the experimenters trained a decoder (Kalman filter) to translate the activity from individual neurons to commands for a virtual arm simulator with realistic dynamics, presented on a computer screen. The participant could control the movements of the arm successfully. Although this work did not describe use of immersive virtual reality, it is an important example of how virtual representations can be used to advance BCI applications; this work served as a test and development platform for the researchers' later work controlling an actual arm using FES [161].

Exoskeletons

Do et al. reported on the control of a robotic system to facilitate walking [215]. A person with paraplegia due to an SCI (T6, i.e., at sixth level of the thoracic spinal cord) was able to walk on a treadmill using the robot while being partially off-loaded by a harness. Imagined gait activated the robot. The imagined movements were identified using a linear Bayesian classifier applied to spectral features from 64-active-channel EEG. The system allowed the user to walk during five, 5-min trials.

More recently, López-Larraz conducted a proof-of-concept study in which four individuals with paraplegia (L1, T11, and T12), resulting from a traumatic SCI, controlled an exoskeleton to facilitate walking without body-weight support [216]. The study participants, who had a good prognosis for gait rehabilitation, attempted movement of the right leg to activate a two-step sequence (right and left) of the exoskeleton. Their system extracted ERD and MRCP from 13 EEG channels (frontocentral, central, and centroparietal; FC3, FCz, FC4, C3, C1, Cz, C2, C4, CP3, CP1,

CPz, CP2, and CP4) and classified them using sparse discriminant analysis (SDA). The authors described the paper as a groundwork for the system's eventual testing as part of a rehabilitation intervention to restore gait in individuals with SCI.

BCI-Triggered Functional Electrical Stimulation

Another fundamental approach to restoring movement is the control of FES devices using BCI technology. As described in Chapter 2, FES is a technique that produces contractions of muscles using highly controlled electrical discharges. Selecting the muscles to stimulate carefully and the sequence in which they are stimulated, results in movements that can be used functionally (e.g., lift a cup from a table).

Pfurtscheller et al. presented an FES system for grasping using EEG activity recorded from a person with tetraplegia (C5) consequence of a complete SCI [217] for the first time. Similar to the researcher's previous work with a hand orthosis [212], the participant imagined foot movements that increased activity in the beta (17Hz) frequency range. The EEG setup used one bipolar EEG channel with the electrodes placed 2.5 cm anterior and posterior to Cz of the 10–20 electrode placement system. Linear discriminant analysis identified the presence of increased beta activity and triggered the FES system. The stimulation facilitated hand function (hand opening and lateral grasp). More specifically, each activation of the BCI would transition onto a different state of the stimulator, which always followed the same sequence: hand opening (i.e., finger extension), finger flexion, thumb flexion, hand opening, and idle (i.e., all stimulation stopped and ready for the next set of BCI activations).

The same group later reported on the control of an implanted FES system to restore upper limb function using a BCI [218]. An LDA classifier used power decreases (ERD) in two frequency bands within the beta frequency range (12–14 Hz and 18–22 Hz). Successful ERD classification triggered a state transition of the FES system, which, like in their previous work [217], included hand opening, performing lateral grasp (i.e., finger flexion followed by thumb flexion), and hand opening. Importantly, control of the FES system was achieved after only three days, suggesting the technology's suitability for use in a clinical setting.

In a demonstration work, Márquez-Chin et al. presented the control of a neuroprosthesis for grasping which was triggered by online classification of ECoG signals [171]. The single participant of that study was a man with a SCI at the C6 level who was fitted with an electrical stimulation system that could classify ECoG activity associated with different movements (wrist flexion, reaching to the right or left). The ECoG signals were recorded previously from a different participant during a monitoring procedure as part of epilepsy treatment. Correct classification of an ECoG recording would trigger a specific stimulation sequence.

Rohm and colleagues demonstrated a hybrid system to restore reaching and grasping functions [219]. In their work, electrical stimulation facilitated hand (lateral grasp and hand opening) and elbow extension. Simultaneously, an actuated orthosis produced elbow flexion movements (for which electrical stimulation was inadequate) with an electrical elbow lock. The participant's shoulder position determined both the opening and closing of the hand and position of the elbow joint. The BCI, which classified ERD patterns elicited by imagined hand and elbow movements using LDA, determined if the voluntary shoulder movements controlled the facilitation of hand or elbow movements or entered a resting state. The system was refined and tested with the help of an individual with a complete SCI at the C4 level resulting in tetraplegia. He could perform functional tasks with the system, such as eating and signing, which he was otherwise unable to do.

More recently, Ajiboye and his colleagues demonstrated intracortical control of an arm support and a multi-channel functional electrical stimulation system to restore reach and grasp. The participant had chronic tetraplegia resulting from a traumatic high-level SCI (C4) sustained eight years before the study [161]. The 53-year-old-man was implanted with two 96-channel microelectrode arrays in the representation of the hand of the primary motor cortex. Four months later, the participant received an implanted FES system to facilitate upper limb movements. For the study, the participant also relied on an actuated support to move his arm against gravity. The BCI used the rate of threshold crossing of individual neurons and the power in high-frequency bands (200 Hz–3 kHz) with both parameters estimated over 20 ms windows. The neural features were processed by a linear filter that translated them into control commands for movement of the elbow, wrist, and hand, or to control the actuated arm support position. After several months of training using different strategies (including control of a virtual arm), the participant could feed himself and drink coffee independently by performing reaching and grasping movement with the system.

In the same year, Friedenberg and colleagues created a new BCI-controlled FES system that allowed direct control of graded muscle contractions, instead of triggering pre-recorded stimulation sequences [220]. The single participant of that study had an SCI (C5–C6) and was implanted with a 96-channel intracortical array placed in the left primary cortex. The neural activity was quantified using mean wavelet power, which was transformed into commands for the 140-channel FES system with support vector regression. Stimulation produced wrist movements, and the neural activity controlled its intensity. With this system, the participant was able to flex his wrist to specific target angles while working against resistance.

CHAPTER 5

The Intersection of Brain–Computer Interfaces and Neurorehabilitation

One of the most exciting applications of BCI systems is the restoration of voluntary movement after paralysis. In contrast to its use as an assistive device, which enhances function by continuous usage, restorative applications use technology for a limited time, as part of a short-term therapeutic intervention. In this context, patients take part in several sessions, after which the use of the BCI stops. Integration of a BCI into neurorehabilitation is motivated by a desire to amplify therapeutic effects, which may translate into greater independence and quality of life. In this chapter we describe several important examples of the use of BCI for restoration of voluntary motor function, providing a sample of the variety of approaches that have been explored to date.

5.1 THE INTERCONNECTION BETWEEN BCI TECHNOLOGY AND NEUROREHABILITATION

Arguably, a critical step in developing BCI technology that led to its use for neurorehabilitation was the control of devices designed to facilitate movement after paralysis. In addition to exploring a new application of BCI systems, the control of motor-restoration devices with brain signals made it possible to provide proprioceptive and somatosensory feedback to the user, in addition to the more common visual modality used in earlier BCI work. The new multisensory feedback provided an opportunity to create systems that could, for the first time, manipulate sensorimotor activity with the potential to have long-term therapeutic effects [221].

The operation of BCI systems that require imagined movements and motor neurorehabilitation overlap in multiple ways. BCIs are, by themselves, rehabilitation devices [222]. Central to the intersection of these two fields is the fact that learning how to operate a BCI using motor imagery and recovering the ability to perform specific motor tasks are both dependent on neuroplastic changes that lead to improved performance (whether this is the control of an electronic device or opening a hand after paralysis).

BCI technology also makes it possible to verify that patients are engaged in therapy; the BCI provides an opportunity, for both the therapists and patients, to ensure that a movement practiced during therapy is being actively imagined or attempted as indicated by the patients' brain activity.

5.1.1 MOTOR IMAGERY

In addition to its extensive use in the context of BCI implementation and control, motor imagery (MI) is a strategy used for neurorehabilitation. As the name suggests, in MI, patients imagine performing precise movements successfully without overt motor output. During neurorehabilitation, patients imagine a specific task repeatedly (often referred to as *Mental Practice*). MI activates similar neural networks as those engaged during the actual execution of a task [223]. Also, this strategy requires little to no resources and can be practiced virtually everywhere. All these factors, and the notion that repeatedly practicing specific tasks produces neuroplastic changes resulting in the recovery of motor function after stroke [224], have led to the adoption of MI into neurorehabilitation. There is good evidence that MI practiced by itself or with other forms of therapy (i.e., constraint-induced movement therapy) may help upper limb function after stroke [225].

5.2 APPROACHES TO INTEGRATION OF BCI TECHNOLOGY INTO NEUROREHABILITATION

To date, there have been two main approaches to integrating BCI systems into the rehabilitation of voluntary movement after stroke and SCI. The first approach consists of using the BCI as an adjuvant to therapy, while the second strategy uses the BCI as part of the primary therapeutic intervention.

5.2.1 BCI AS AN ADJUVANT TO THERAPY

The use of a BCI as an adjuvant to therapy requires that patients are actively participating in therapy during their rehabilitation. In addition, they also undergo supplementary sessions in which they learn to operate a BCI using motor imagery. The motivation for this approach is the notion that when patients learn to produce motor-related brain activity that resembles that of healthy individuals, the entire remaining neural structures involved in the production of movement will also become normalized and, in turn, increase the effects of therapy. One clear advantage of this approach is that it can be combined with any intervention Among the vast repertoire of therapy modalities available during the rehabilitation of an individual with a neurological condition.

5.2.2 BCI FOR TRIGGERING VOLUNTARY MOVEMENT DURING REHABILITATION

The second approach to integrating a BCI into rehabilitation consists of using the technology to trigger an additional device designed to facilitate movement. This approach is motivated by the possibility of ensuring the motor-related activity produced while imagining or attempting movements is met with the somatosensory and proprioceptive information that is temporally relevant and congruent with the practiced movement. This concept is illustrated in Figure 5.1.

Figure 5.1: **A BCI triggers a device for movement restoration.** This figure illustrates the use of a BCI to trigger a functional electrical stimulation system that allows the user to hold a cup. In this conceptual image, the BCI detects the user's intention or attempt to grasp a cup and triggers electrical stimulation designed specifically to assist the required grasping movement. In contrast to the concepts illustrated in Figure 5.1, the resulting feedback (orange arrow) is not only visual, but it also includes somatosensory and proprioceptive information resulting from the contact with the cup as well as the contracting muscles and movement. This simultaneous presence of efferent (i.e., motor command) and afferent (i.e., sensory feedback) activity is believed to lead to neuroplastic changes responsible for recovering voluntary movement.

5.3   MOTOR RESTORATION AFTER STROKE USING BCI TECHNOLOGY

Individuals who have had a stroke may benefit significantly from the development of BCI-based rehabilitation interventions.

5.3.1   UPPER LIMB FUNCTION

One of the areas that have seen the most activity in the exploration of BCI technology for motor rehabilitation is the restoration of upper limb function after stroke. Whether used in combination with virtual representations of a limb, an actuated orthosis, a rehabilitation robot, or an electrical stimulation system, the work in this area has provided the first opportunity to understand the advantages and challenges of using a BCI as a neurorehabilitation tool. The next few sections describe important examples of BCI systems integrated with rehabilitation for restoring reaching and grasping.

Arm Function

In 2014, Ang and colleagues [226] reported on the combination of a BCI and a rehabilitation robot to restore arm reaching function in individuals with hemiparesis resulting from a stroke (Figure 5.2). In their randomized control trial, 11 participants received a 12-session intervention in which assistance from a Manus robot (Interactive Motion Technologies USA, Watertown, MA) was triggered by imagining reaching toward 8 targets displayed radially on a screen using their affected limb. The remaining 14 completed the treatment using the robot system alone. Each session was 90 min long, with 20 min dedicated to set up the equipment and a targeted 160 trials. The study demonstrated that the intervention was effective and safe. Also, the improvements recorded when using both interventions were comparable even though the number of repetitions per session using the BCI and robot (136), was only a fraction of when the robot was used alone (1,040). Clinical and technical details for this study can be found in Tables 5.1 and 5.2.

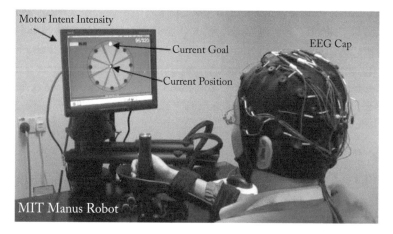

Figure 5.2: BCI-controlled rehabilitation robot for restoring arm movement. Image modified from [227].

Table 5.1: Clinical details of the sample study exploring the use of BCI to restore arm function after stroke.

Reference	Intervention	Imagined, Attempted, Observed Movement	Rehabilitation Goal	Number of Sessions	Primary Functional Outcome Measure	Increase in Primary Outcome
[226]	BCI-controlled rehabilitation robot	Imagined reaching toward radially placed targets	Arm reaching	12	Fugl-Meyer Assessment (FMA)	4.5

Table 5.2: Technical details of the sample study exploring the use of BCI to restore arm function after stroke.

Reference	Brain Activity Recording Technique	Number of Sensors or Electrodes	Brain Activity Feature	Classifier or Detector
[226]	EEG	27	Spatio-frequency filtered EEG	Naive Bayesian Parzen window classification algorithm

Hand Function

In 2009, Daly and colleagues reported using a BCI-triggered FES system for restoring hand movements [228]. The participant of that study was a 46-year-old woman who sustained a stroke ten months before the study. As a result, she was unable to move her affected fingers in isolation. The experimenters recorded EEG signals from the ipsilesional cortex during either attempted or imagined finger movements followed by relaxation. The FES, designed to facilitate extension of the index finger, was triggered whenever the power in a specific frequency range decreased below a moving average calculated over multiple trials. After only nine sessions, the participant was able to move her index finger isolated from the rest. Clinical and technical details for this study can be found in Tables 5.3 and 5.4.

In 2013, Mihara et al. conducted a proof-of-principle study to test the use of NIRS to promote recovery of hand function [229]. The study recruited individuals who had sustained a stroke within the previous six months. Before introducing the BCI system, each individual received daily one-hour physical, occupational, and speech (if needed) therapy sessions (maximum total of three hours) for one week. Ten participants underwent 20-min mental practice sessions for 2 weeks. Each session included 10 min of a motor-imagery-controlled video game and 10 min of kinesthetic motor imagery training. Neurofeedback, representing imagined hand movements, was presented to the patients on a computer screen throughout the session. The remaining (ten) participants underwent a similar procedure that differed only because the biofeedback activation was random. The real biofeedback reduced the level of impairment more than the random one.

Li and colleagues explored the efficacy of a motor-imagery BCI to restore hand function on eight individuals with severe hemiplegia resulting from a stroke no more than six months before their participation in the study [230]. For eight weeks, participants received conventional therapy (physical and exercise therapy and acupuncture) and BCI-triggered electrical stimulation therapy three times per week in sessions lasting 60–90 min (Figure 5.3). Before the intervention, participants underwent training to help them perform motor imagery using the task of "imagining drinking water". During each session, participants engaged in several games controlled by imagined movements of the affected or unaffected hand. After five successful activations of the BCI, the system triggered electrical stimulation designed to facilitate hand opening. A control group, which included 7 participants, received conventional therapy sessions and 20 min of FES therapy alone (i.e., without a BCI). Individuals who underwent the BCI-controlled FES intervention displayed improved motor function, activation of both cerebral hemispheres, and enhanced ERD.

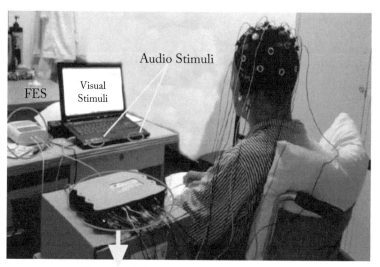

Figure 5.3: BCI-controlled rehabilitation functional electrical stimulation system for restoring hand movement. Image modified from [230].

Ono et al. reported a study describing the influence of feedback on the efficacy of hand function restoration using a BCI-controlled hand orthosis in subacute and chronic stages of rehabilitation [231]. Six of the 12 participants of that pilot study imagined movements with their affected hand. The resulting ERD would trigger an actuated orthosis designed to facilitate finger extension. The remaining participants received visual feedback in the form of a virtual hand opening. After 12–20 sessions, the participants who received feedback with the orthosis displayed increased EMG activity and improvement in the finger function scores of the Stroke Impairment Assessment Set (SIAS) [232], suggesting the importance of somatosensory feedback for the efficacy of an intervention.

In another study, Ang et al. reported the combination of a BCI with a robot designed to assist hand opening and closing as well as pronation and supination of the forearm [233] (Figure 5.4). In that three-arm, single-blind randomized control trial, 21 participants took part in 18 x 90-min sessions focused on wrist, grasp, and release functions that could be delivered using robotic-assisted hand therapy, with or without BCI activation, or by a therapist. Participants allocated to the robotic therapy completed 120 trials, followed by 30 min of treatment provided by a therapist. The BCI detected ERD using 27 EEG channels. Despite not finding inter-group differences, participants who received therapy using a BCI-activated robot showed a significantly greater improvement in impairment on three out of the four measurements performed throughout the study.

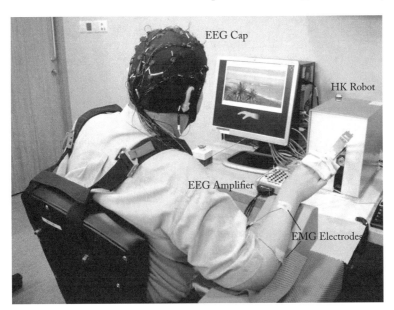

Figure 5.4: BCI-controlled rehabilitation robot for restoring wrist and hand movements. Image modified from [233].

Pichiorri and colleagues conducted a pilot randomized control trial to measure the efficacy of a low-cost clinically-viable motor imagery intervention assisted with BCI technology [234]. The study included 28 individuals with subacute stroke who participated in the research while still at the hospital. In addition to the treatment for stroke delivered as part of their hospital-based rehabilitation, participants also received three weekly sessions for one month. They practiced imagining grasping and finger extension (hand opening) with their affected hand for approximately 30 min. A BCI system guided the MI for 14 patients, by projecting virtual hands on a white blanket placed over their real hands (Figure 5.5). Simultaneously, the treating therapist could also monitor the motor-imagery-related activity and guide the patients accordingly. The remaining 14 individuals received feedback on a screen with similar hand images as the ones used by the treatment group but the BCI did not control the feedback. After the intervention, the participants who underwent MI training with the BCI improved their impairment level more than the control group and had a higher probability of experiencing clinically meaningful improvements. Also, the reductions in power associated with the imagined movements in the affected hemisphere became more robust for the BCI group, indicating neuroplastic changes.

Figure 5.5: BCI-controlled-guided system facilitating motor imagery of hand movements. Image modified from [234].

More recently, Kim and his colleagues conducted a randomized control trial to test the efficacy of a BCI-controlled FES system, combined with action observation training (AOT), to restore the upper limb function. The participants were in the first year of living with hemiplegia after sustaining a stroke [235]. AOT, in which patients observe another person perform goal-directed actions, has been introduced as a supplement to stroke rehabilitation for restoring movement after stroke; AOT activates cortical areas similar to those involved in the actual execution of a movement.

All of the 32 participants received daily 30-min sessions of occupational therapy, and 17 of them also took part in three weekly sessions of the same duration, which included the combined AOT with BCI+FES. A screen showed multiple actions that included several activities used in daily life (e.g., folding a towel, using scissors, turning a faucet, among many others). Patients were asked to focus mainly on the wrist movements while watching the videos, which, when done correctly, triggered the BCI, subsequently activating the stimulation designed to produce wrist extension. After four weeks of training, participants who received AOT combined with BCI+FEST showed more significant reductions in impairment and improvements in hand function.

Biasiucci et al. also reported the therapeutic effects of combining BCI technology with FES [236]. In their study, individuals with moderate to severe hemiplegia at least ten months after sustaining a stroke attempted to extend the wrist and hand (i.e., hand opening). For 14 participants, this attempt was identified by a BCI, which in turn triggered FES designed to produce the intended movements. The research team delivered stimulation at random to a control group of 13 participants. Both groups had two therapy sessions per week for five weeks. At the end of the intervention, participants in the BCI group demonstrated a decrease in impairment that was clinically meaningful, and the strength in the stimulated muscle (*extensor digitorum communis*) was significantly higher than for the control group. Notably, the study participants experienced significant recovery of voluntary motor function, which lasted a year after the intervention. Detailed EEG analysis revealed that the connectivity within affected motor areas increased. This denser connectivity and motor improvements were correlated.

Table 5.3: Clinical details of the sample studies exploring the use of BCI to restore hand function after stroke

Reference	Intervention	Imagined, Attempted, Observed Movement	Rehabilitation Goal	Sessions	Primary Functional Outcome Measure	Increase in Primary Outcome
[228]	BCI-Triggered FES	Attempted and imagined index finger extensions and relaxation	Index finger extension	9	Angle of index finger extension	26°
[229]	Neurofeedback presented on a screen	Videogame-guided motor imagery of elbow and fingers followed by 10 min of kinesthetic motor imagery with neurofeedback	Hand movement	6	Fugl-Meyer Assessment (FMA)	6.6
[230]	BCI-controlled FES using videogame-based biofeedback to guide imagined upper limb movements	Wrist and hand extension	Wrist and hand extension	24	Fugl-Meyer Assessment (FMA)	9 points (approx.)
[231]	BCI-controlled hand orthosis and neurofeedback presented on a screen in the form of a virtual hand	Attempted finger extension	Finger extension	20-Dec	Stroke Impairment Assessment Set (SIAS)	1
[233]	BCI-controlled rehabilitation robot	Wrist rotation and supination; grasp and release	Hand and finger movement	18	Fugl-Meyer Assessment (FMA)	9.7
[234]	Anatomically congruent BCI-controlled biofeedback presented on a screen covering the participants' hands	Grasping and finger extension	Hand function	12	Fugl-Meyer Assessment (FMA)	44 ± 34.7
[235]	Action observational training combined with BCI-controlled FES	Observed wrist movements from a video recording	Wrist extension	20	Fugl-Meyer Assessment (FMA) –Upper Extremity sub score	7.87
[236]	BCI-triggered FES	Attempted hand opening	Wrist and finger extension	10	Fugl-Meyer Assessment (FMA) –Upper Extremity sub score	6.6 ± 3.6

Table 5.4: Technical details of the sample studies exploring the use of BCI to restore hand function after stroke

Reference	Brain Activity Recording Technique	Number of Sensors or Electrodes	Brain Activity Feature	Classifier or Detector
[228]	EEG	1	Reduction in power	Threshold comparison
[229]	NIRS	3	β-coefficient and t values for the task-related cortical activation change	
[230]	EEG	16	Spatio-temporo-spectral features	Support vector machine (SVM)
[231]	EEG	1 bipolar channel		
[233]	EEG	27	Spatio-frequency filtered EEG	Naive Bayesian Parzen window classification algorithm
[234]	EEG	31	Power decrease in spatio-spectral filtered EEG	
[235]	EEG	2	Summation of sensory motor rhythm (SMR; 12–15 Hz; unfocused attention) and mid-beta rhythm (16–20 Hz; focused attention), divided by theta band activity (4–7 Hz)	Threshold comparison
[236]	EEG	16	Sensorimotor rhythms in the 4–30 Hz range	Gaussian classifier

Hand and Arm Function

A study conducted by Ramos-Murguialday et al. in 2013, compared the effects of using hand and arm orthoses, activated using a BCI or randomly, followed by one hour of physical therapy [237] (Figure 5.6). All participants had lived for at least ten months with no active finger flexion due to a stroke. They attempted reaching, grasping, retrieving, and releasing an apple with their affected hand. For 16 of the 32 individuals, the ERD produced by the attempted movement was detected by a BCI, which triggered a hand or arm actuated orthosis. In comparison, the orthoses were activated at random for the remaining 16 participants. Both groups received one hour of physical therapy immediately after the orthoses sessions. After 17 days of training, the BCI group showed increased EMG activity and improvements in the upper limb components of the Fugl-Meyer Assessment impairment measure. Clinical and technical details for this study can be found in Tables 5.5 and 5.6. A recent follow-up study conducted by the same researchers, in which both participant groups followed an in-home treatment program, revealed that the significant changes observed using the BCI-based therapy intervention lasted six months after the original intervention [238].

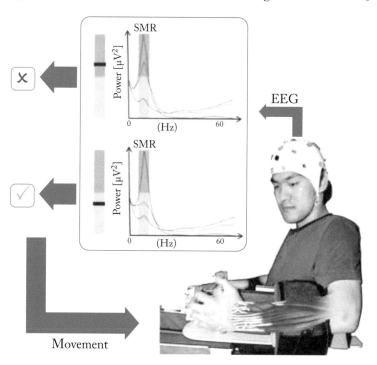

Figure 5.6: **BCI-controlled orthosis for restoring upper limb movement.** Image modified from [237].

Table 5.5: Clinical details of the sample study exploring the use of BCI to restore hand and arm function after stroke

Reference	Intervention	Imagined, Attempted, Observed Movement	Rehabilitation Goal	Sessions	Primary Functional Outcome Measure	Increase in Primary Outcome
[237]	Physiotherapy immediately following BCI-controlled reaching or grasping orthosis session	Intended reach and grasp movements of the affected limb	Arm reaching and hand opening and closing	17.8 ± 14 days	Combined hand and modified arm Fugl-Meyer Assessment (cFMA)	3.41 ± 0.563

Table 5.6: Technical details of the sample study exploring the use of BCI to restore hand and arm function after stroke

Reference	Brain Activity Recording Technique	Number of Sensors or Electrodes	Brain Activity Feature	Classifier or Detector
[237]	EEG	16	Power decrease in spatio-spectral filtered EEG	-

5.3.2 LOWER LIMB FUNCTION

A recent study by Mrachacz-Kersting measured the effects of using BCI technology to restore ankle function [192]. Their work included 22 participants who had sustained a stroke a minimum of 6 months earlier. All of them attempted dorsiflexion movements of their affected foot. Thirteen of the participants underwent an intervention in which an MRCP-based (Section 4.5.2) BCI identified the attempted movement and triggered a single stimulation pulse applied to the peroneal nerve (Figure 5.7). Notably, the researchers controlled the stimulation very carefully so that that the resulting sensory activity would match precisely the negative peak of the MRCP (indicative of the maximum activation of the motor cortex). The remaining nine participants received random stimulation. After only three sessions, the results revealed improved impairment scores, walking speed, and foot-tapping frequency for the BCI group. Similarly, the corticospinal tract excitability to the tibialis anterior muscle, responsible for performing foot dorsiflexion, increased for the treatment group but not for the control group. Clinical and technical details for this study can be found in Tables 5.7 and 5.8. The proof-of-concept study highlighted the importance of timing (i.e., synchronization between efferent and afferent activity) to produce neuroplastic changes leading to motor recovery.

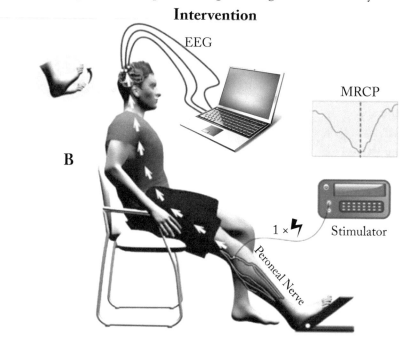

Figure 5.7: BCI-triggered peroneal stimulation for restoring ankle function. Image modified from [192].

Table 5.7: Clinical details of the sample study exploring the use of BCI to restore lower limb function after stroke

Reference	Intervention	Imagined, Attempted, Observed Movement	Rehabilitation Goal	Sessions	Primary Functional Outcome Measure	Increase in Primary Outcome
[192]	Single electrical stimulation of the tibialis anterior muscle, timed so that resulting afferent activity corresponded with peak negativity of the MRCP	Attempted foot dorsiflexion	Foot dorsiflexion	3	Fugl-Meyer motor assessment (FM)	0.8 ± 0.46

Table 5.8: Technical details of the sample study exploring the use of BCI to restore lower limb function after stroke

Reference	Brain Activity Recording Technique	Number of Sensors or Electrodes	Brain Activity Feature	Classifier or Detector
[192]	EEG	10	Motor-related cortical potential (MRCP)	Locality Preserving Projections (LPP) learning algorithm combined with a linear discriminant analysis (LDA) classifier

5.4 MOTOR RESTORATION AFTER SPINAL CORD INJURY USING BCI-TECHNOLOGY

Compared to the research conducted to restore voluntary movement after stroke, the number of studies exploring the BCI technology's efficacy for movement rehabilitation is minimal. The limited number of research reports may be due to the size of the population of individuals with SCI, or to the often-greater severity of the impairments associated with this condition. Additionally, rehabilitation after SCI frequently requires interventions for both the right and left sides of the body. However, restoring the ability to move voluntarily after SCI is essential. Improvements in motor function can lead not only to an increased quality of life but to full and successful reintegration into society.

5.4.1 UPPER LIMB FUNCTION

As in the case of rehabilitation after stroke, the use of BCI+FEST for the restoration of upper limb function has been primarily focused on recovery of hand function (i.e., grasping) and often excludes reaching function. After SCI, individuals living with tetraplegia often list restoration of upper limb function as a top rehabilitation priority [73, 239].

Hand Function

Osuagwu and colleagues [240] tested the efficacy of a BCI-triggered FES intervention to restore grasping function in individuals with tetraplegia resulting from SCI (C4–C7, i.e., between the cervical 4 and 7 levels of the spinal cord). The intervention consisted of 20, one-hour sessions delivered 3–5 times a week to 12 individuals in the subacute (<3 months) stage of rehabilitation.

Seven participants received BCI+FEST and the remaining five received FEST alone. The BCI was designed to discriminate between attempted movement and no movement using LDA. The system used three bipolar EEG channels which recorded activity over the sensory motor cortex bilaterally and medially. When the BCI detected the intention to move, it triggered the FES system. With the exception of one person in the FEST group, all of the assessed participants (five and three from the treatment and control groups, respectively) experienced an increased range of motion (ROM) after the intervention in both wrists. The FEST+BCI group experienced a statistically significant bilateral improvement in all muscle groups defined by the experiments which included those involved in control of shoulder, upper arm, lower arm, and wrist and finger flexion and extension. In comparison, significant improvements in the control group were only observed on the muscle group involved in shoulders (left and right) and those of the right hand. Clinical and technical details for this study can be found in Tables 5.9 and 5.10.

Table 5.9: Clinical details of the sample study exploring the use of BCI to restore hand function after SCI resulting in tetraplegia

Reference	Intervention	Imagined, Attempted, Observed Movement	Rehabilitation Goal	Sessions	Primary Functional Outcome Measure	Increase in Primary Outcome
[240]	BCI-Triggered FES	Attempted hand movements	Hand opening and closing	20	Range of motion of left and right wrists	Right hand = 15.3°, Left hand = 16.8°

Table 5.10: Technical details of the sample study exploring the use of BCI to restore hand function after SCI resulting in tetraplegia

Reference	Brain Activity Recording Technique	Number of Sensors or Electrodes	Brain Activity Feature	Classifier or Detector
[240]	EEG	3 bipolar channels	Power in the 7–30 Hz EEG band	Linear discriminant analysis (LDA)

5.4.2   LOWER LIMB FUNCTION

As in the case of rehabilitation after stroke, and despite the technical and clinical challenges associated with restoration of the ability to walk after SCI, research has started applying BCI technology to improve gait.

In a recent study, eight individuals with chronic paraplegia resulting from SCI (7 complete and one incomplete) underwent a 12-month intervention to restore the ability to walk [241]. In that long-term intervention, Donati et al. tested the combination of multiple technologies to provide sensory feedback using a BCI. The intervention, which combined physical therapy, incorporated: (1) an avatar in an immersive virtual reality environment with visuo-tactile feedback controlled by the BCI and operated both seated and with the help of a standing frame; (2) training using a body-weight support system on both a treadmill and on the ground; and (3) walking with an exoskeleton—capable of providing somatosensory feedback— controlled by BCI. Tactile feedback was delivered on the participants' forearms, and it corresponded to the rolling of the feet. Participants first trained to operate the BCI using imagined arm movements and later learned how to control it by imagining moving their legs. All participants experienced an improved sensation in several dermatomes and regained the ability to control voluntary muscles below the level of injury, which translated into a better ability to walk. Four of the participants with complete SCI were reclassified as incomplete injuries after the intervention. Clinical and technical details for this study can be found in Tables 5.11 and 5.12.

Table 5.11: Clinical details of the sample study exploring the use of BCI to restore gait after SCI resulting in paraplegia

Reference	Intervention	Imagined, Attempted, Observed Movement	Rehabilitation Goal	Sessions	Primary Functional Outcome Measure	Increase in Primary Outcome
[241]	(1) BCI-controlled avatar in immersive reality space while seated, (2) BCI-controlled avatar in immersive reality space while standing, (3) walking with a body weight support (BWS) robotic system, (4) walking with BWS system on overground track, (5) walking using a BCI-controlled BWS system on a treadmill, (6) walking with an exoskeleton enhanced to deliver tactile feedback	Imagined walking	Gait restoration	2,052 (1,958 hours)	-	-

Table 5.12: Clinical details of the sample study exploring the use of BCI to restore gait after SCI resulting in paraplegia

Reference	Brain Activity Recording Technique	Number of Sensors or Electrodes	Brain Activity Feature	Classifier or Detector
[241]	EEG	16	–	Linear Discriminant Analysis

CHAPTER 6

Implementation of a BCI-Triggered Functional Electrical Stimulation Therapy

6.1 INTRODUCTION

As seen in various clinical trials previously introduced, one promising approach for integrating BCI with neurorehabilitation consists of its use with FEST. As explained in Chapters 1 and 2, FEST (functional electrical stimulation therapy) is an intervention in which patients practice specific functional and purposeful tasks assisted by electrical stimulation. Briefly, in a typical FEST session, a patient is asked to perform a functional movement and, after a few seconds (10–20 [96]), a therapist activates electrical stimulation explicitly designed to assist the targeted functional movement. The number and complexity of movements usually change over time, as the ability to perform voluntary movements improves. Finally, electrical stimulation stops at the end of the therapy (i.e., the stimulation is used as a short-term intervention). This therapeutic benefit of electrical stimulation has been documented since its initial use during the 1960s [40, 67, 242]. FEST has resulted in some of the largest motor function improvements after stroke [243] and has also been more effective when compared to state-of-the-art physical and occupational therapy in the population of individuals with SCI [96]. FEST has become an important tool for therapists specialized in the restoration of voluntary movement after stroke and SCI.

Despite the critical results achieved with FEST, which have allowed a successful return to daily activities for many individuals with paralysis, research continues in the rehabilitation community to develop new interventions with ever-increasing efficacy. One population that remains a significant challenge to rehabilitate consists of individuals in the chronic stages of rehabilitation (i.e., individuals who have lived with paralysis for a year or more with minimal improvement in function since injury/disease onset). In this case, there is a generalized understanding that the likelihood of recovery resulting from an intervention is limited. Also, the rehabilitation options are minimal when the impairment is severe as most rehabilitation interventions require that patients retain a minimal amount of residual [244].

One of the advantages of FEST is that it can be used in chronic individuals with a very high level of impairment (i.e., when the ability to move is very limited or absent). However, in these cases, determining when to trigger the stimulation may be challenging as this often relies on the experience of the treating therapists to identify the intention to move using indirect indicators. As described in previous chapters, one of the potential mechanisms that may explain FEST's efficacy is the presence of anatomically and temporally congruent efferent (motor) and afferent (sensory) activity. Consequently, the difficulty in determining when to trigger the stimulation may directly impact the efficacy of the intervention.

6.2   ENHANCING FEST WITH BCI TECHNOLOGY

One of the potential strategies to enhance the efficacy of FEST may be to use it in combination with BCI. This combination of neurotechnologies makes it possible to use neurological indicators of the intention to move and trigger the stimulation, potentially increasing the likelihood that a motor command is met with the corresponding sensory feedback, leading to neuroplastic changes that improve voluntary motor function. During the last several years, our group has developed techniques to combine BCI and FEST. In this chapter, we describe some of our experiences and contributions.

6.2.1   GENERAL DESCRIPTION OF THE SYSTEM DESIGN

The BCI+FEST that we have created combines the experience gained over two decades, developing and testing FEST technologies and interventions as well as our initial work in the BCI field, which used four-contact subdural electrodes, conducted during the early 2000s. The clinical experience-driven approach has resulted in a system for supporting FEST delivery in both research and clinical environments. The BCI works as a detector that monitors the brain's activity for decreases of power that occur as patients attempt different movements under the guidance of a therapist.

At this time, our BCI+FEST system requires one treating therapist and one person overlooking the BCI operation. The therapist guides the session and is responsible for the configuration and operation of the FES system. In turn, the BCI operator monitors the BCI state and can adjust, if necessary, its parameter values to modify its behavior.

Important aspects of BCI for FEST in our system

The system's design supports using it in a clinical setting, outside of a research laboratory environment, and integrating it with FEST. This section describes the critical features that we implemented to achieve this clinical suitability.

Unintrusive BCI

We wanted to make the BCI as unintrusive as possible to both the patient and the actual therapy. From the perspective of the person receiving the therapy, the operation of the BCI does not require any training; patients attempt to perform a movement (i.e., guided by the therapist during rehabilitation), and the BCI detects the intention to move. The BCI subsequently triggers the electrical stimulation to produce the practiced movement. Additionally, the system hides the BCI's actual activation; there is no visual or auditory feedback for the patient. We removed this feature when it became evident that visual/auditory feedback often became a distraction from the therapy; for some patients, the focus of the therapy became to receive the visual or auditory confirmation that the BCI was working instead of the careful and mindful execution of the practiced movements. Only one small visual indicator consisting of a light emitting diode (LED) is provided for the therapist. This LED reflects the BCI state and can be mounted anywhere where it is deemed convenient (Figure 6.1).

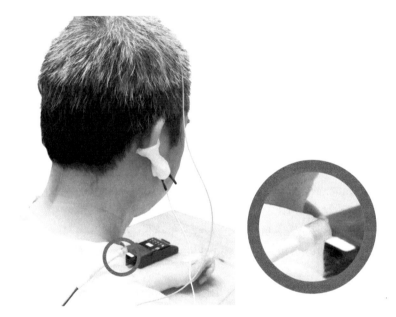

Figure 6.1: The visual indicator used by the therapist to identify the state of the BCI remains out of the patients' line of sight.

The Therapist Can Override the BCI

Another feature is the possibility to override the BCI at any moment during therapy. This mechanism allows therapists to use their clinical judgement to activate the electrical stimulation if they consider that the BCI missed detecting the attempted movement. In addition to making it possible to incorporate the experience that a therapist may bring to the delivery of the therapy, this feature minimizes frustration for the patient.

BCI Parameters Can Be Modified At Any Moment

A critical challenge for integrating BCI into neurorehabilitation, particularly when combined with FEST, is that this intervention often includes practicing numerous movements that often change throughout rehabilitation according to the patient's goals and progress. Besides, the movements can be complex, involving multiple joints and phases. Although different for each individual, more often than not, rehabilitation of movement of the upper limb will rarely incorporate, exclusively, a single arm or hand movement. Instead, therapy will often include different reaching types (e.g., forward or lateral reaching, reaching to the mouth, or reaching to the opposite shoulder) as well as different grasping styles (e.g., palmar or lateral grasp). Examples of the movements typically used during BCI+FEST are displayed in Table 6.1. More importantly, these arm and hand movements are combined into full synergies (i.e., reaching and grasping) that allow practicing functional movements. This variety of movements, coupled with the unpredictability of if or when they will be used, makes using advanced artificial intelligence techniques difficult. This problem is made worse by the fact that each patient is unique and that the amount of time available in a clinical setting would likely not allow the collection of sufficient data for creating a model capable of dealing with all of the potential cases. These reasons were the motivation for choosing to implement the BCI as a detector (rather than a classifier).

The BCI uses a power threshold and a temporal threshold to identify reductions in power in subject-specific frequency bands, signaling an attempt to move. Both thresholds need to be reached for the BCI to activate. The BCI detects the intention to move when the power of an incoming (and processed) EEG signal falls below the power threshold. Additionally, the BCI is only activated if the decreased power in the incoming signal is sustained for the amount of time specified by the temporal threshold.

The BCI operator can adjust both the power and temporal thresholds to modify the BCI's responsiveness at any moment during a therapy session. The system also has the optional capacity to self-adjust the power threshold using information collected over the most recent trials. This simple implementation makes it possible to adjust the system quickly to the changing conditions resulting from different practiced movements throughout the therapeutic intervention.

Table 6.1: Examples of movements practiced during upper limb rehabilitation along with the muscles that require stimulation for executing them

Stimulated Muscle	Movement Produced	Sample Practiced Movements
Anterior deltoid	Shoulder flexion	Forward reaching
	Shoulder adduction	Reaching for a ball
	Medial arm rotation	
Middle deltoid	Shoulder abduction	Lateral reaching
Posterior deltoid	Shoulder extension	Retrieving the arm
	Shoulder abduction	Pulling an object back
	Lateral arm rotation	
Biceps brachii	Elbow flexion	Holding a cup while using a straw
	Shoulder flexion	
	Shoulder abduction	
Triceps	Elbow extension	Returning a utensil to a table
	Shoulder extension	
Flexor digitorum superficialis/ profundus	Finger flexion	Holding a mug
	Wrist flexion	
Extensor digitorum	Finger abduction	Releasing a ball
	Finger extension	Opening the hand before grasping and object
	Wrist extension	
Lumbricals	Finger flexion	Holding a ball
Second dorsal interosseous	Finger flexion	Holding a book
	Finger abduction	
Opponens pollicis	Thumb opposition	Grasping a water bottle
Abductor pollicis brevis	Thumb abduction	Opening hand to grasp an object

Minimized Setup Time

Another essential consideration when designing the BCI is the time required to set it up during each session. The typical length of one of our sessions is one hour, a value comparable to the one used by the clinical program in our institution. To increase our system's clinical viability, we wanted to ensure that the setup could be completed as quickly as possible to maximize the actual therapy time in each session. For this reason, the activity of the brain is captured with a single monopolar channel only, using the ear lobes as ground and reference. Early BCI research included single-channel designs with both EEG [183, 212] and subdural signals [170]. Although this design makes the recording of EEG activity more susceptible to noise and makes the use of spatial filtering techniques impossible, it does allow for a quick setup that can be repeated with multiple individuals throughout a single day, as it takes place in a rehabilitation setting. The average time required to set up the BCI system, measured over 130 1-hour sessions, is 11 min, and it is comparable to the time required for setting up the FES system.

6.2.2   IMPLEMENTATION

Hardware Implementation

The BCI+FEST system consists of a biopotential amplifier (QP511, Grass Telefunken, Germany) used to acquire the single EEG channel, a data acquisition system (USB-6363, National Instruments, U.S.) and custom-made software written in LabView (National Instruments, U.S.A.), which processes the EEG signal in real time and generates a synchronization signal (pulse) whenever a decrease of power is detected (see BCI Parameters Can be Modified at any Moment). The same data acquisition system used to digitize the EEG activity converts the synchronization signal into a transistor-transistor logic (TTL) pulse. This pulse is sent to an industrial-grade isolator and subsequently connected to the Compex Motion electrical stimulator [245].

Software Implementation

EEG Processing

The digitized EEG activity is band-limited between 0.05–40 Hz. The root mean square (RMS) is then calculated after squaring the filtered signal. A moving average is calculated using 1 sec of this power estimate. The resulting signal is then multiplied by a gain which determines the gradient slope, providing a level of control between the speed of the transition between low and high power states. This signal is then compared, constantly, with a power threshold (BCI Parameters Can be Modified at any Moment). When the power in the signal falls below this threshold, a timer is triggered, which quantifies the duration of the reduction of power. If the power reduction lasts for a set value (temporal threshold; BCI Parameters Can be Modified at any Moment), and the output of the BCI is enabled (details below), the BCI generates a pulse to trigger the stimulation. The user can adjust both the power and temporal thresholds at any moment. Figure 6.2 displays a flowchart describing implementation of the BCI.

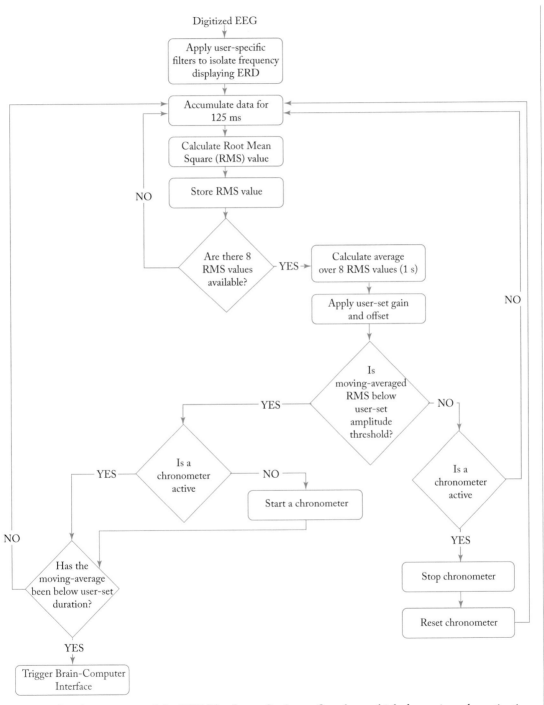

Figure 6.2: Implementation of the BCI. The figure displays a flowchart which determines the activation of the BCI.

Stimulation Protocols

The Compex Motion is a four-channel programmable stimulator, making it possible to design stimulation sequences to produce different movements. Its proprietary software allows defining when each channel is activated and at what intensity. For example, a stimulation sequence may include providing stimulation to muscles responsible for extending the wrist and fingers, necessary to open a hand, stopping the stimulation, and immediately activating a third channel that produces finger flexion to close the hand. Notably, the system also makes it possible to determine the transition between the different stimulation stages (e.g., switching between finger extension and flexion). The transition can be performed automatically, using a timer, or triggered by an external input, such as a digital or analog sensor. It is important to emphasize that the stimulation sequences are defined by the person's clinical needs undergoing rehabilitation and developed under the guidance of the treating therapist. In our system, the transition between the different stages in the stimulation sequences occurs using an external electronic pulse. This pulse is provided by either the BCI or an external switch, which the therapist can use to command the FES system, bypassing the BCI. Figures 6.3 and 6.4 provides examples of some of our stimulation sequences, including the transition of stimulation states with the BCI. The figures show how the manual switch ends the stimulation sequence, returning both the stimulator and the BCI to an idle state (referred to as unarmed in the user interface). In this idle state, activations of the BCI will not affect the stimulation. Activation of the manual button once again will reactivate the BCI control, and any new activation will trigger the stimulation sequences (Figure 6.2).

6.2.3   OPERATION

Calibration of the System

The BCI requires calibration before using it. The procedure is very similar to that used by other systems that use motor-related activity. With patients seated in front of a screen, they are presented with four cues which they must follow:

1. a "READY" signal, which serves as an indicator that a trial has started;

2. a visual indicator displaying a movement (e.g., a hand performing a specific type of grasp);

3. a "GO" signal that indicates to the patients that they should attempt or execute the movement specified in Step 2; and

4. a "STOP" signal which tells the patients that the trial is over.

These steps are repeated multiple times to collect a minimum of 80 trials. If the therapy is likely to include rehabilitation of both the left and right limbs, as it is common in the SCI population, the process is repeated for each side.

In addition to recording EMG activity (if present) of the muscles involved in the movement, our configuration process also records every experimental cue as well as EEG activity from eight locations of the extended 10–20 system: F3, F4, Fz, C3, C1, C4, C2, Cz, with reference and ground on the mastoid processes. Each trial is aligned with the "GO" signal and split into 12-sec-long segments after capturing the data. These segments are then used to generate ERD maps to identify the electrodes and frequency bands that display a reduction in power. We use this information to implement the BCI for detecting the intention to move of both the right and left upper limbs.

Use of the System During Therapy

After setting up the FES and BCI systems in each session, the therapists will ask the person to prepare to execute a specific movement, previously discussed with the patient. The therapist will then press the manual button once, which will arm the BCI so that any activation will trigger a change in the stimulation sequence. The button will also turn on a visual indicator (see Unintrusive BCI) which the therapist can use to confirm that the BCI+FEST system is ready. The BCI detects the patient's attempted movement, activates the stimulation corresponding to the first stage in the stimulation sequence (e.g., hand opening and reaching forward), and sets its state to unarmed (i.e., further BCI activations will not result in change in the stimulation). Once the patient completes the movement assisted by the stimulation, the therapist will cue the patient to perform the next phase of the movement and press the switch again which will arm the BCI. Once again, the BCI will detect the attempted movement which will trigger the next phase of the stimulation (e.g., hand closing) and disengage the BCI control (unarm). The same cycle is repeated for all the stages of the stimulation sequence (in the case of the examples, the next two stages could involve retrieving the extended arm followed by opening the hand). After the final stimulation stage, the therapist activates the switch one more time, ending the stimulation altogether. This additional button activation is needed because the patient attempts no movement at the end of the stimulation sequence, when all channels are deactivated. Figures 6.3 and 6.4 show two stimulation sequences with different stages. At any moment, the therapist can use the switch to activate the stimulation bypassing the BCI.

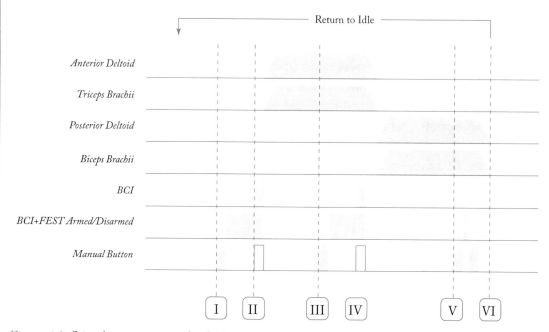

Figure 6.3: Stimulation sequence for facilitating forward reaching using the BCI+FEST system. Each manual button or BCI activation results in a change in stimulation state (extend, retrieve, and idle). Starting from an idle state and with the patient relaxed: (I) the therapist activates a manual button "arming" the BCI, making it possible for it to command the electrical stimulator. At the same time, the therapist asks the patient to attempt reaching forward; (II) when the BCI detects the patient's attempt to reach forward, it activates the stimulation on the *anterior deltoid* and *triceps brachii* muscles which facilitates the reaching function. The system is also "unarmed" disrupting the BCI control of the FES device. This feature ensures that subsequent unintended BCI activations do not affect the stimulation. Please note that the therapist can also activate the stimulation if necessary (represented by a blank rectangle); (III) the therapist arms the BCI again and indicates to the patient to retrieve the arm; (IV) the attempt to perform the movement is identified by the BCI which then interrupts the stimulation of the *triceps brachii* and *anterior deltoid*. The stimulator program then triggers the stimulation of the *biceps brachii* and *posterior deltoid* automatically. This stimulation assists the patient retrieve the arm; (V) once the movement is completed, the therapist presses the manual button twice which stops all stimulation; and (VI) returns the BCI+FEST system to an idle state, ready to start a new trial.

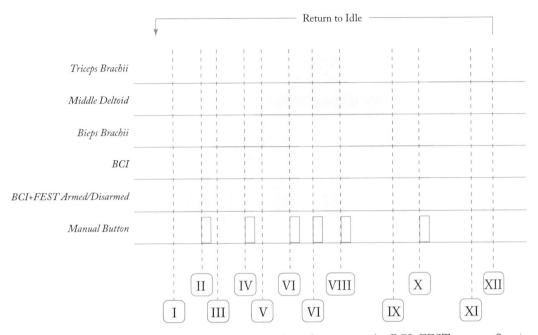

Figure 6.4: Stimulation sequence for facilitating lateral reaching using the BCI+FEST system. Starting with the patient's arm relaxed: (I) the therapist activates the manual button once "arming" the system to allow the BCI to control the electrical stimulator. The therapist also asks the patient to attempt flexing the elbow to raise the forearm; (II) the BCI detects the attempt to move and triggers the stimulation to the *biceps brachii* muscle. The system automatically disarms the system to interrupt the BCI control of the FES device; (III) the therapist then arms the system again by pressing the manual button and asks the person to lift the elbow laterally (elbow abduction); (IV) upon detecting the attempt to move, the BCI triggers stimulation to the *anterior* and *posterior deltoid* muscles, facilitating the attempted movement. The system is disarmed automatically; (V) once again, the therapist arms the system, and asks the patient to perform extend the arm (elbow extension); (VI) as with the previous steps, the BCI detects the attempted movement and discontinues stimulation to the *biceps brachii* muscle and initiates stimulation of the *triceps brachii* muscle. The stimulation to the *anterior* and *posterior deltoid* muscles is maintained and the system is disarmed; (VII) the therapist arms the system by pressing the manual button and asks the patient to perform elbow flexion, reversing the last movement; (VIII) the attempted movement results in stimulation to the *triceps brachii* stopped and stimulation of the *biceps brachii* muscle active. This is followed by an automatic stopping of stimulation to the *anterior* and *posterior deltoid* muscle which results in lowering of the elbow (shoulder adduction); (X) in the final stimulation phase the therapists ask the patient to extend the elbow to return the arm to the starting position; (X) after detecting the attempted movement, the BCI stops the stimulation to the *biceps brachii* muscle and activates stimulation to the *triceps brachii* muscle, resulting elbow extension; (XI) finally, the therapist presses the manual button twice rapidly which stops all stimulation; and (XII) returns the BCI+FEST system to an idle state, ready to start a new trial.

6.2.4   CLINICAL TESTING

Restoration of Upper Limb Function in Chronic Severe Hemiplegia

One of the motivations for developing the BCI+FEST system was to enhance the efficacy of FEST. Specifically, we noticed through our other studies that some individuals with chronic severe hemiplegia would not respond to the intervention while others improved their ability to use their arms and hands after FEST. Individuals with chronic severe hemiplegia represent a population that is often best suited to participate in BCI+FEST therapy as most interventions, except FEST, require that patients retain residual motor function. Also, the likelihood of recovery decreases with time, although this long held belief is gradually changing. We considered that one potential factor for the observed lack of response was the difficulty in determining when patients were attempting the movement during therapy. With this in mind, we identified BCI technology as a potential way to confirm if and when patients were attempting the movement and consequently the suitable moment to trigger the stimulation.

Our initial tests included two individuals with chronic severe hemiplegia for whom conventional FEST did not produce any change in their upper limb function [246, 247]. Both participants had sustained a stroke six years before they received the BCI+FEST intervention. Every intervention that they had tried previously had failed to improve their ability to use their arm and/or hand. The first participant, a 64-year-old man, received 40 sessions (90-min/session), delivered three times a week, to restore his ability to reach with his left hand after sustaining a hemorrhagic stroke. At the start of the therapy, he was completely unable to use his arm, which presented as flaccid. During the treatment, the FES system was programmed to assist with five reaching movements including reaching forward, reaching to the opposite knee, reaching to the mouth, reaching to the opposite shoulder, and reaching sideways (lateral reaching). The BCI triggered each of the stimulation phases (i.e., reaching and retrieving). At the end of the intervention, the participant experienced a significant and clinically meaningful improvement in impairment [246].

The second case was a 57-year-old male who had sustained a thrombotic stroke resulting in severe left hemiplegia. As with the first participant, he received therapy thrice a week. However, the sessions lasted only 60 min, and the intervention included 80 sessions. During the intervention, numerous movements were practiced, adjusted by the therapist according to his progress. The initial sessions emphasized reaching movements, and later hand function was added, focusing on hand opening. Later, functional movements (i.e., reaching and grasping a bottle) were included in the sessions and changed regularly. The ability to reach and grasp improved dramatically for the participant, which allowed him to perform relevant and meaningful tasks again since sustaining the stroke [247].

Restoration of Reaching and Grasping in Subacute Spinal Cord Injury

As mentioned earlier in the book, FES therapy trials in our laboratory have produced promising results in both the stroke and SCI populations especially when applied early on after injury, i.e., in the subacute phase-post injury. Whereas we did see positive changes with conventional FES in chronic stages post-recovery as well, the results were not as consistent, which motivated us to integrate BCI with FES. As discussed in the previous subsection, from a technical perspective we were able to successfully integrate BCI with FES in stroke and were able to retrain complex upper extremity movements in this population. Although we have only done this as case studies in chronic stroke, the results are nonetheless promising. This, and the fact that individuals with SCI, especially those with a higher level and severity of injury face tremendous challenges with respect to independence in activities of daily living, was our incentive to trial our BCI+FES therapy in SCI. As far as we know, our group is among the first ones to trial a dual channel BCI system in combination with FES to retrain bilateral complex upper extremity movements that involve components of reaching and grasping in subacute as well as chronic SCI. In our preliminary work, we recruited five and three individuals in the subacute and chronic stages of rehabilitation, respectively. Our early results suggest that BCI+FES can be successfully applied in this population, as was evident from feasibility data. More than half of the study participants in the subacute group showed clinically significant improvements in upper extremity function, as measured by standard SCI upper extremity outcome tools. With respect to the three chronic participants, one showed significant improvement which translated, for example, in the person being able to drink water from a bottle independently, which he was not able to do prior to BCI+FES intervention. The improvements in the other two participants were more subtle. These early results indicate that, whereas there is merit to the use of this technology in the rehabilitation of individuals with SCI, we are just starting to be able to identify the "right" patients for whom this intervention can produce life-changing results.

CHAPTER   7

Concluding Remarks

7.1   FINAL THOUGHTS

In this book, we present a broad perspective on the potential implementation of neuromodulation in rehabilitation. We look closer at FES and BCI, as the combination of these two technologies provides an opportunity to enhance the engagement of both motor and sensory pathways, and support integration during re-training of voluntary movement. However, the perspectives and approaches can be used in various neuromodulation venues.

There is growing evidence that BCI devices and peripheral stimulation can be coupled. Currently, several methods are under study but not clinically viable. Ultimately, before new technologies become safe and feasible for clinical practice, there is a trial-and-error period in humans, which needs to occur in each disease separately. Thus, the state of technology and its role in clinical intervention will always be an evolutionary process, requiring multiple revisions and refinements. It is important to note that disease and presentation of disease also evolve, and neurorehabilitation must adapt to the patient's changing needs.

Electrical stimulation technology itself has seen significant refinement and a multitude of applications over the years. In the domain of sensory function restoration, it leads the fields of cardiology, ophthalmology, urology, and many more. In the domain of motor function restoration, today, FEST is widely accepted in North America as a clinical tool to retrain upper extremity function, walking, balance etc., post neurological injuries. Although there are several hypotheses related to the physiological mechanisms behind the improvements seen, in our opinion, the improvements are a result of its ability to stimulate the neuromuscular system below the level of injury to retrain the nervous system to recover a specific motor task. Combining BCI and FES allows delivering stimulation below the level of injury with accurate timing of the motor execution of the task, thus strengthening the afferent-efferent pathways. Here, we discussed several BCI+FES systems; ours can be triggered using minimal electrodes, reducing the setup time, which is critical for clinical implementation. This is a step forward in making BCI accessible at a clinical level and to a point where it can be more easily coupled with existing or evolving technologies.

Although at this time BCI remains a laboratory-based tool, given the attention it is receiving from the research community and the general public, it is currently being implemented in clinical studies conducted in clinical settings. Thus, we are hopeful that before too long it will be available as a viable rehabilitation tool.

7.2   A REFLECTION ON THE STATUS TO DATE

While some BCI concepts are not feasible to apply in clinical settings, it is possible to incorporate some set-ups into a neuromodulatory program with a therapist. These BCI applications are most promising in that they can be used to treat patients and perhaps have an improved outcome in setting efferent and afferent pathway development during rehabilitation.

7.3   WHAT THE FUTURE MAY BRING

While there will always be laboratory applications for BCI devices, a significant impact will be achieved when it can further the gains realized from neuroplasticity, thereby enhancing patient outcomes. Currently, we know several neuro-modulatory methods to be neuro-restorative. BCI technology has the potential to enhance this effect.

From a neurorehabilitation perspective, several important questions will need addressing as the BCI field evolves. Some of them include:

When is the best time to deliver a BCI-based therapy?

One of the most exciting applications made possible with BCI is treating individuals in the chronic stages of rehabilitation effectively. Several studies have suggested the efficacy of the intervention in individuals many years after acquiring a neurological condition. Individuals with chronic severe hemiplegia have shown considerable improvements in function when treated with a combination of BCI+FES therapy. However, as is the case with many forms of therapy, further work is still necessary to determine if there is a time to maximize the intervention's effects. For example, the time elapsed between sustaining a stroke and receiving care is critical for recovery. One potential new use to BCI technology is to engage stroke survivors in a motor imagery-based intervention early in the acute stage once medically stable.

Which populations can benefit from including a BCI in rehabilitation?

Another essential unknown is who is appropriate for receiving therapy that incorporates BCI technology. Although the initial research and development of BCI targeted locked-in individuals, the use of BCI for neurorehabilitation has expanded to other populations that can benefit from this technology by including people with less severe forms of impairment. However, the question remains if there is a benefit of using BCI technology to treat individuals with moderate to mild impairments. It may be difficult to justify the control of a device for motor restoration using a BCI in individuals who have some residual function and are able to initiate the targeted movements; in our own experience, the BCI has interfered with the treatment once the ability to move voluntarily has improved past a certain point. However, just like athletes who engage in mental practice to improve motor skills, it may be possible to create new BCI paradigms explicitly designed to take advantage of similar approaches for rehabilitating individuals with higher voluntary function.

What would be the therapeutic effect of identifying specific movements?

As stated earlier in the book, a critical area of work in the BCI, or more specifically in the BMI, field is the identification of specific imagined or attempted movements. Intracortical recordings have produced the most successful examples of this technology. However, a non-invasive alternative would allow exploring therapeutic effects of using a device that can identify specific intended movements and facilitate them on an individual with paralysis (i.e., through FES or robotic system). In most interventional studies, the practiced movements are previously selected by the therapist or the technical team behind the BCI implementation. BCI combined with a system capable of restoring multiple complex movements that are not restricted to preprogrammed synergies, may provide a new level of diversity and richness in the movements practiced during therapy. The enhanced flexibility could allow for an unprecedented opportunity to incorporate tasks selected according to the patient's goals. For now, methods similar to those that have been demonstrated in the context of hybrid systems combining robotic and FES devices, can likely be used to simulate the identification of specific movements by integrating environmental information (e.g., through RFID or computer vision techniques) and allow an initial exploration on the human-machine interaction requirements and therapeutic effects of this future technology.

Can individuals with cognitive impairments participate in BCI-based therapy?

Another future research area consists of using BCI technology to deliver therapy for motor restoration in individuals with cognitive impairments. Stroke can result in cognitive deficits, which may reduce the ability to solve problems and process or understand language. BCI technology provides a unique opportunity to verify that a person receiving therapy is engaged in the practiced task. As such, individuals with difficulty comprehending (i.e., receptive aphasia) may benefit from a BCI by increasing the level of certainty on the patient's engagement during therapy.

7.4 THE ROLE OF BCI TECHNOLOGY IN NEUROREHABILITATION

The use of BCI technology for neurorehabilitation is in its infancy and it will likely undergo significant changes and numerous iterations prior to its adoption in the clinical rehabilitation setting. One of the key areas that remains is exploring new avenues where the BCI can produce transformational changes in individuals with long standing neurological ailments.

7.4.1 COMBINATION OF INTERVENTIONS

BCI technology shows great potential as an adjunct to existing or evolving rehabilitation therapies. It is important to remember that the integration of BCI and FES, and BCI and robots discussed in this book (both from a technical and clinical perspectives) are only two examples of integrating technologies, each with previous applications in neurorehabilitation. Besides, our understanding of the mechanisms associated with neuroplasticity and how to enhance them for neurorehabilitation is constantly evolving. This new knowledge will continue to lead to new pharmacological, biological and electrophysiological techniques to augment the neuroplastic effects of therapy, many of which will be translated and implemented using engineering tools [221]. The future will likely bring new combinations of procedures and technologies, including BCIs, to increase the effects of neurorehabilitation.

7.4.2 FEEDBACK DEVICE FOR THE TREATING CLINICIAN

BCI technology can also be envisioned as a monitoring tool in the context of neurorehabilitation, where the therapists can use BCI to monitor the status of the nervous system during the delivery of rehabilitation. For example, with suitable development of analysis paired with predictive capabilities, it may one day be possible to assess and predict an individual's responsiveness to a specific intervention. In turn, a treating therapist may use this information to monitor and optimize therapy such that it remains challenging and in turn maximizes gains.

Initial evidence of this type of application was reported recently. A team of researchers explored the relationship between the EEG oscillatory activity, used for BCI implementation, and recovery of voluntary upper limb function [248]. Their research included data collected from 22 patients with chronic severe impairments resulting from stroke and revealed that the ERD (8–12 Hz) in the ipsilesional hemisphere that grew progressively stronger with rehabilitation was correlated with better rehabilitation outcomes. Other research groups have observed similar correlations between brain activity and clinical rehabilitation measures [249].

7.4.3   REHABILITATION OF OTHER CONDITIONS

Although this book focused on the use of BCI technology for the rehabilitation of voluntary movement, there are other important areas of rehabilitation that may benefit from these devices. It is likely that in the next few years, BCIs will be used to address other neurological conditions or impairments in addition to motor restoration. Of particular relevance is the potential of BCI-based interventions to address cognitive impairments as well as emotion and mood regulation deficits, all common after stroke and with a significant negative impact in rehabilitation. More specifically, post-stroke cognitive impairments can be expressed as a decreased ability to sustain attention, slow information processing, memory deficits, reduced semantic fluency, and difficulties generating or processing speech. These deficits can directly impact a person's ability to participate in rehabilitation as most interventions require a minimum ability to understand and follow instructions during rehabilitation. Similarly, post-stroke depression, of which two main characteristics are lack of motivation and action, can affect participation in therapy directly. The last decade has seen an increase in BCI-based neurofeedback interventions to address both cognitive [250–254] and emotion [255] impairments after stroke. Importantly, it has been suggested that BCI operation during rehabilitation alone may serve as a tool for motivating patients to participate in rehabilitation [256]. In other populations, BCI-based neurofeedback interventions have shown promising results for the management of attention-deficit hyperactive disorder (ADHD) [257–260], as well as mild cognitive impairment in older individuals [261].

Another emerging application of BCI technology is the restoration of somatosensory function. In this context, a BCI system is used to deliver microstimulation to the cerebral cortex eliciting a tactile sensation [262, 263]. Although intracortical microsimulation may be better suited for long-term applications (e.g., neurally-controlled prosthetic devices), the previously mentioned studies suggest the possibility of using BCI devices to address pain problems associated with, for example stroke, SCI, and amputation [264].

7.5   RECOMMENDED READING

Although the book presented representative references describing the use of BCI systems, the following are recommended for their historical or clinical relevance.

With respect to movement rehabilitation:

- [234] provides a unique example of the use of a BCI as an adjuvant to therapy, focusing on the normalization of motor-related cortical activity as a technique to enhance the effects of therapy.

- [228] presents one of the earliest examples of the clinical use of FES controlled by a BCI to restore hand movements in a person with a stroke.

- In a more recent study, [236] provides the results of some of the largest clinical trials conducted to date combining FES and BCI systems for the rehabilitation of hand movements in the chronic stroke population with long lasting effects [236].

- [226] represents early work on the rehabilitation of arm movements by combining robot-assisted therapy and a BCI system.

- [237] describes a clinical study for restoring both hand and arm function, using a BCI and orthotic systems as an adjuvant to therapy.

- [192] provides a unique perspective on the opportunity to use BCI technology to synchronize efferent and afferent activity with high-precision, leading to extraordinary neuroplastic-changes associated with motor performance.

With respect to rehabilitation of cognitive functions:

- The work presented in [251] explores the effects of EEG-based neurofeedback for improving working memory and verbal shot and long-term memory in individuals who had sustained a stroke.

Concerning rehabilitation of somatosensory function:

- [264] showcases initial work on the application of BCI techniques and technologies to reduce musculoskeletal pain.

References

[1]   Lackland DT, Roccella EJ, Deutsch AF, Fornage M, George MG, Howard G, et al. Factors influencing the decline in stroke mortality: A statement from the American Heart Association/American Stroke Association. *Stroke*. 2014 Jan;45(1):315–53. DOI: 10.1161/01.str.0000437068.30550.cf. 1

[2]   Johnson W, Onuma O, Owolabi M, Sachdev S. Stroke: A global response is needed. *Bulletin of the World Health Organization*. 2016 Sep;94(9):634–634A. DOI: 10.2471/ BLT.16.181636. 1

[3]   Adams RD, Ropper AH, Victor M. *Principles of Neurology*. 6th ed. New York: McGraw-Hill, Health Professions Division; 1997. 2, 5

[4]   Marx JA, Hockberger RS, Walls RM, Adams J, Rosen P. *Rosen's Emergency Medicine : Concepts and Clinical Practice*. 6th ed. Philadelphia, PA: Mosby/Elsevier; 2006. 2

[5]   Ojaghihaghighi S, Vahdati SS, Mikaeilpour A, Ramouz A. Comparison of neurological clinical manifestation in patients with hemorrhagic and ischemic stroke. *World Journal of Emergency Medicine*. 2017;8(1):34–8. DOI: 10.5847/wjem.j.1920-8642.2017.01.006. 2

[6]   Vrocher D. *Annals of Emergency Medicine*. 2004;44(6):675–6. 2

[7]   Krueger H, Koot J, Hall RE, O'Callaghan C, Bayley M, Corbett D. Prevalence of individuals experiencing the effects of stroke in Canada: Trends and Projections. *Stroke*. 2015 Aug;46(8):2226–31. DOI: 10.1161/STROKEAHA.115.009616. 2

[8]   Goeree R, Blackhouse G, Petrovic R, Salama S. Cost of stroke in Canada: A 1-year prospective study. *Journal of Medical Economics*. 2005 Jan;8(1-4):147–67. DOI: 10.3111/200508147167. 2

[9]   Virani Salim S., Alonso Alvaro, Benjamin Emelia J., Bittencourt Marcio S., Callaway Clifton W., Carson April P., et al. Heart disease and stroke statistics update: A report from the American Heart Association. *Circulation*. 2020 Mar;141(9):e139–596. 2

[10]   Stack CA, Cole JW. Ischemic stroke in young adults. *Current Opinion in Cardiology*. 2018 Nov;33(6):594–604. DOI: 10.1097/HCO.0000000000000564. 3

[11]   Oyinbo CA. Secondary injury mechanisms in traumatic spinal cord injury: A nugget of this multiply cascade. *Acta Neurobiologiae Experimentalis*. 2011;71(2):281–99. 4

[12]   Kalsi-Ryan S, Karadimas SK, Fehlings MG. Cervical spondylotic myelopathy: The clinical phenomenon and the current pathobiology of an increasingly prevalent and devastating disorder. *The Neuroscientist: A Review Journal Bringing Neurobiology, Neurology and Psychiatry*. 2013 Aug;19(4):409–21. DOI: 10.1177/1073858412467377. 4

[13]   Nouri A, Tetreault L, Singh A, Karadimas SK, Fehlings MG. Degenerative cervical myelopathy: epidemiology, genetics, and pathogenesis. *Spine*. 2015 Jun;40(12):E675–693. DOI: 10.1097/BRS.0000000000000913. 4

[14]   Wilson JR, Fehlings MG, Kalsi-Ryan S, Shamji MF, Tetreault LA, Rhee JM, et al. Diagnosis, heritability, and outcome assessment in cervical myelopathy: A consensus statement. *Spine*. 2013 Oct;38(22 Suppl 1):S76–77. DOI: 10.1097/BRS.0b013e3182a7f4bf. 4

[15]   Jain NB, Ayers GD, Peterson EN, Harris MB, Morse L, O'Connor KC, et al. Traumatic spinal cord injury in the United States, 1993-2012. *JAMA*. 2015 Jun;313(22):2236–43. DOI: 10.1001/jama.2015.6250. 4

[16]   Lasfargues JE, Custis D, Morrone F, Carswell J, Nguyen T. A model for estimating spinal cord injury prevalence in the United States. *Paraplegia*. 1995 Feb;33(2):62–8. DOI: 10.1038/sc.1995.16. 4

[17]   Spinal Cord Injury (SCI) 2016 Facts and figures at a glance. *The Journal of Spinal Cord Medicine*. 2016 Jul;39(4):493–4. DOI: 10.1080/10790268.2016.1210925. 4

[18]   Noonan VK, Fingas M, Farry A, Baxter D, Singh A, Fehlings MG, et al. Incidence and prevalence of spinal cord injury in Canada: A national perspective. *Neuroepidemiology*. 2012;38(4):219–26. DOI: 10.1159/000336014. 4

[19]   Kirshblum SC, Memmo P, Kim N, Campagnolo D, Millis S. American Spinal Injury Association. Comparison of the revised 2000 American Spinal Injury Association classification standards with the 1996 guidelines. *American Journal of Physical Medicine & Rehabilitation*. 2002 Jul;81(7):502–5. DOI: 10.1097/00002060-200207000-00006. 4

[20]   Fehlings MG, Tetreault LA, Wilson JR, Kwon BK, Burns AS, Martin AR, et al. A clinical practice guideline for the management of acute spinal cord injury: Introduction, rationale, and scope. *Global Spine Journal*. 2017 Sep;7(3 Suppl):84S–94S. DOI: 10.1177/2192568217703387. 5

[21]   Levin MF, Kleim JA, Wolf SL. What do motor "recovery" and "compensation" mean in patients following stroke? *Neurorehabilitation and Neural Repair*. 2009 May;23(4):313–9. DOI: 10.1177/1545968308328727. 5

[22] Dobkin BH. Confounders in rehabilitation trials of task-oriented training: Lessons from the designs of the EXCITE and SCILT multicenter trials. *Neurorehabilitation and Neural Repair*. 2007, 21(1):3–13. DOI: 10.1177/1545968306297329. 5

[23] Teasell R, Salbach NM, Foley N, Mountain A, Cameron JI, Jong A de, et al. Canadian stroke best practice recommendations: Rehabilitation, recovery, and community participation following stroke. Part One: Rehabilitation and recovery following stroke; 6th Edition Update 2019. *International Journal of Stroke: Official Journal of the International Stroke Society*. 2020 Oct;15(7):763–88. DOI: 10.1177/1747493019897843. 5, 43

[24] Harnett A, Bateman A, McIntyre A, Parikh R, Middleton J, Arora M, et al. Spinal cord injury rehabilitation praciticesrehabilitation practices. *Spinal Cord Injury Research Evidence*. https://scireproject.com/evidence/rehabilitation-evidence/rehabilitation-practices/; 2020. 5, 23

[25] Cramer SC, Sur M, Dobkin BH, O'Brien C, Sanger TD, Trojanowski JQ, et al. Harnessing neuroplasticity for clinical applications. *Brain: A Journal of Neurology*. 2011 Jun;134(6):1591–609. DOI: 10.1093/brain/awr039. 6

[26] Kleim JA, Jones TA. Principles of experience-dependent neural plasticity: Implications for rehabilitation after brain damage. *Journal of Speech, Language, and Hearing Research: JSLHR*. 2008 Feb;51(1):S225–239. DOI: 10.1044/1092-4388(2008/018). 6, 7

[27] Marquez-Chin C, Popovic MR. Functional electrical stimulation therapy for restoration of motor function after spinal cord injury and stroke: A review. *BioMedical Engineering OnLine*. 2020 May;19(1):34. DOI: 10.1186/s12938-020-00773-4. 7, 8, 13

[28] Kapadia N, Popovic MR. Functional electrical stimulation therapy for grasping in spinal cord injury: An overview. *Topics in Spinal Cord Injury Rehabilitation*. 2011 Jul;17(1):70–6. DOI: 10.1310/sci1701-70. 7

[29] Kapadia NM, Zivanovic V, Furlan JC, Craven BC, McGillivray C, Popovic MR. Functional electrical stimulation therapy for grasping in traumatic incomplete spinal cord injury: Randomized control trial. *Artificial Organs*. 2011 Mar;35(3):212–6. DOI: 10.1111/j.1525-1594.2011.01216.x. 7, 24

[30] Kapadia N, Masani K, Catharine Craven B, Giangregorio LM, Hitzig SL, Richards K, et al. A randomized trial of functional electrical stimulation for walking in incomplete spinal cord injury: Effects on walking competency. *The Journal of Spinal Cord Medicine*. 2014 Sep;37(5):511–24. DOI: 10.1179/2045772314Y.0000000263. 7, 21, 22

[31] Nagai MK, Marquez-Chin C, Popovic MR. Why is functional electrical stimulation therapy capable of restoring motor function following severe injury to the central nervous

system? In: Tuszynski M, editor. *Translational Neuroscience*. Springer; 2016. pp. 479–98. DOI: 10.1007/978-1-4899-7654-3_25. 8

[32] Ibáñez J, Monge-Pereira E, Molina-Rueda F, Serrano JI, del Castillo MD, Cuesta-Gómez A, et al. Low latency estimation of motor intentions to assist reaching movements along multiple sessions in chronic stroke patients: A feasibility study. *Frontiers in Neuroscience*. 2017;11. DOI: 10.3389/fnins.2017.00126. 9

[33] Hackett ML, Duncan JR, Anderson CS, Broad JB, Bonita R. Health-related quality of life among long-term survivors of stroke : Results from the Auckland Stroke Study, 1991-1992. *Stroke*. 2000 Feb;31(2):440–7. DOI: 10.1161/01.STR.31.2.440. 9

[34] Muralidharan A, Chae J, Taylor DM. Extracting attempted hand movements from EEGs in people with complete hand paralysis following stroke. *Frontiers in Neuroscience*. 2011 Mar;5. DOI: 10.3389/fnins.2011.00039. 9

[35] Kapadia NM, Nagai MK, Zivanovic V, Bernstein J, Woodhouse J, Rumney P, et al. Functional electrical stimulation therapy for recovery of reaching and grasping in severe chronic pediatric stroke patients. *Journal of Child Neurology*. 2014 Apr;29(4):493–9. DOI: 10.1177/0883073813484088. 9, 24

[36] Bertani R, Melegari C, De Cola MC, Bramanti A, Bramanti P, Calabrò RS. Effects of robot-assisted upper limb rehabilitation in stroke patients: A systematic review with meta-analysis. *Neurological Sciences*. 2017 Sep;38(9):1561–9. DOI: 10.1007/s10072-017-2995-5. 9, 36

[37] Gonzalez Andino SL, Herrera-Rincon C, Panetsos F, Grave De Peralta R. Combining BMI stimulation and mathematical modeling for acute stroke recovery and neural repair. *Frontiers in Neuroscience*. 2011;5. DOI: 10.3389/fnins.2011.00087. 9

[38] Kralj A, Vodovnik L. Functional electrical stimulation of the extremities: Part 1. *Journal of Medical Engineering & Technology*. 1977 Jan;1(1):12–5. DOI: 10.3109/03091907709161582. 11

[39] Peckham PH, Knutson JS. Functional electrical stimulation for neuromuscular applications. *Annual Review of Biomedical Engineering*. 2005;7:327–60. DOI: 10.1146/annurev.bioeng.6.040803.140103. 11

[40] Liberson WT. Functional electrotherapy: Stimulation of the peroneal nerve synchronized with the swing phase of the gait of hemiplegic patients. *Archives of Physical Medicine*. 1961;42:101–5. 11, 85

[41] Long C. An electrophysiologic splint for the hand. *Archives of Physical Medicine and Rehabilitation*. 1963 Sep;44:499–503. 11

[42]   Angeli CA, Edgerton VR, Gerasimenko YP, Harkema SJ. Altering spinal cord excitability enables voluntary movements after chronic complete paralysis in humans. *Brain: A Journal of Neurology*. 2014 May;137(Pt 5):1394–409. DOI: 10.1093/brain/awu038. 12

[43]   Mortimer JT, Bhadra N. Peripheral nerve and muscle stimulation. In: *Neuroprosthetics. World Scientific*; 2004. pp. 638–82. (Series on Bioengineering and Biomedical Engineering; Volume 2). DOI: 10.1142/9789812561763_0020. 12

[44]   Ethier C, Oby ER, Bauman MJ, Miller LE. Restoration of grasp following paralysis through brain-controlled stimulation of muscles. *Nature*. 2012 May;485(7398):368–71. DOI: 10.1038/nature10987. 12

[45]   Popovic MR, Keller T, Pappas IP, Dietz V, Morari M. Surface-stimulation technology for grasping and walking neuroprosthesis. *IEEE Engineering In Medicine And Biology Magazine : The Quarterly Magazine of the Engineering in Medicine & Biology Society*. 2001 Jan;20(1):82–93. DOI: 10.1109/51.897831. 13

[46]   Teasell R, Salbach NM, Foley N, Mountain A, Cameron JI, Jong A de, et al. Canadian stroke best practice recommendations:Rehabilitation, recovery, and community participation following Stroke. Part One: Rehabilitation and recovery following stroke; 6th Edition Update 2019. *International Journal of Stroke: Official Journal of the International Stroke Society*. 2020 Jan;1747493019897843. DOI: 10.1177/1747493019897843. 13

[47]   Kalra L, Langhorne P. Facilitating recovery: Evidence for organized stroke care. *Journal of Rehabilitation Medicine*. 2007 Mar;39(2):97–102. DOI: 10.2340/16501977-0043. 13

[48]   Prvu Bettger JA, Stineman MG. Effectiveness of multidisciplinary rehabilitation services in postacute care: State-of-the-science. A review. *Archives of Physical Medicine and Rehabilitation*. 2007 Nov;88(11):1526–34. DOI: 10.1016/j.apmr.2007.06.768. 13

[49]   Seiffge DJ, Werring DJ, Paciaroni M, Dawson J, Warach S, Milling TJ, et al. Timing of anticoagulation after recent ischaemic stroke in patients with atrial fibrillation. *The Lancet Neurology*. 2019 Jan;18(1):117–26. DOI: 10.1016/S1474-4422(18)30356-9. 13

[50]   Stroke Unit Trialists' Collaboration. Organised inpatient (stroke unit) care for stroke. *The Cochrane Database of Systematic Reviews*. 2013 Sep;(9):CD000197. 13

[51]   Anonymous. 1. Initial Stroke Rehabilitation Assessment. Canadian Stroke Best Practices. https://www.strokebestpractices.ca/en/recommendations/stroke-rehabilitation/initial-stroke-rehabilitation-assessment/; 2016. 13, 17, 20

[52]   Sheffler LR, Taylor PN, Gunzler DD, Buurke JH, Ijzerman MJ, Chae J. Randomized controlled trial of surface peroneal nerve stimulation for motor relearning in lower limb

hemiparesis. *Archives of Physical Medicine and Rehabilitation*. 2013 Jun;94(6):1007–14. DOI: 10.1016/j.apmr.2013.01.024. 14

[53] Jaqueline da Cunha M, Rech KD, Salazar AP, Pagnussat AS. Functional electrical stimulation of the peroneal nerve improves post-stroke gait speed when combined with physiotherapy. A systematic review and meta-analysis. *Annals of Physical and Rehabilitation Medicine*. 2020 May. DOI: 10.1016/j.rehab.2020.03.012. 14

[54] Stanic U, Acimović-Janezic R, Gros N, Trnkoczy A, Bajd T, Kljajić M. Multichannel electrical stimulation for correction of hemiplegic gait. Methodology and preliminary results. *Scandinavian Journal of Rehabilitation Medicine*. 1978;10(2):75–92. 14

[55] Bogataj U, Gros N, Malezic M, Kelih B, Kljajić M, Acimović R. Restoration of gait during two to three weeks of therapy with multichannel electrical stimulation. *Physical Therapy*. 1989 May;69(5):319–27. DOI: 10.1093/ptj/69.5.319. 14

[56] Daly JJ, Roenigk K, Holcomb J, Rogers JM, Butler K, Gansen J, et al. A randomized controlled trial of functional neuromuscular stimulation in chronic stroke subjects. *Stroke*. 2006 Jan;37(1):172–8. DOI: 10.1161/01.STR.0000195129.95220.77. 14

[57] Tan Z, Liu H, Yan T, Jin D, He X, Zheng X, et al. The effectiveness of functional electrical stimulation based on a normal gait pattern on subjects with early stroke: A randomized controlled trial. *BioMed Research International*. 2014;2014:545408. DOI: 10.1155/2014/545408. 14

[58] Kesar TM, Reisman DS, Perumal R, Jancosko AM, Higginson JS, Rudolph KS, et al. Combined effects of fast treadmill walking and functional electrical stimulation on post-stroke gait. *Gait & Posture*. 2011 Feb;33(2):309–13. DOI: 10.1016/j.gaitpost.2010.11.019. 14

[59] Sauder NR, Meyer AJ, Allen JL, Ting LH, Kesar TM, Fregly BJ. Computational design of fastFES treatment to improve propulsive force symmetry during post-stroke gait: A feasibility study. *Frontiers in Neurorobotics*. 2019;13:80. DOI: 10.3389/fnbot.2019.00080. 14

[60] Kafri M, Laufer Y. Therapeutic effects of functional electrical stimulation on gait in individuals post-stroke. *Annals of Biomedical Engineering*. 2015 Feb;43(2):451–66. DOI: 10.1007/s10439-014-1148-8. 15

[61] Burridge JH, Taylor PN, Hagan SA, Wood DE, Swain ID. The effects of common peroneal stimulation on the effort and speed of walking: A randomized controlled trial with chronic hemiplegic patients. *Clinical Rehabilitation*. 1997 Aug;11(3):201–10. DOI: 10.1177/026921559701100303. 15

[62] Taylor P, Burridge J, Dunkerley A, Wood D, Norton J, Singleton C, et al. Clinical audit of 5 years provision of the Odstock dropped foot stimulator. *Artificial Organs*. 1999 May;23(5):440–2. DOI: 10.1046/j.1525-1594.1999.06374.x. 15

[63] Taylor PN, Burridge JH, Dunkerley AL, Wood DE, Norton JA, Singleton C, et al. Clinical use of the Odstock dropped foot stimulator: Its effect on the speed and effort of walking. *Archives of Physical Medicine and Rehabilitation*. 1999 Dec;80(12):1577–83. DOI: 10.1016/S0003-9993(99)90333-7. 15

[64] Street T, Singleton C. A clinically meaningful training effect in walking speed using functional electrical stimulation for motor-incomplete spinal cord injury. *The Journal of Spinal Cord Medicine*. 2018 May;41(3):361–6. DOI: 10.1080/10790268.2017.1392106. 15

[65] Hausdorff JM, Ring H. Effects of a new radio Frequency-Controlled neuroprosthesis on gait symmetry and rhythmicity in patients with chronic hemiparesis. *American Journal of Physical Medicine & Rehabilitation*. 2008 Jan;87(1):4–13. DOI: 10.1097/PHM.0b013e31815e6680. 16

[66] Stein RB, Everaert DG, Thompson AK, Chong SL, Whittaker M, Robertson J, et al. Long-term therapeutic and orthotic effects of a foot drop stimulator on walking performance in progressive and nonprogressive neurological disorders. *Neurorehabilitation and Neural Repair*. 2010 Feb;24(2):152–67. DOI: 10.1177/1545968309347681. 16

[67] Vodovnik L, Kralj A, Stanic U, Acimovic R, Gros N. Recent applications of functional electrical stimulation to stroke patients in Ljubljana. *Clinical Orthopaedics and Related Research*. 1978 Mar;(131):64–70. DOI: 10.1097/00003086-197803000-00009. 17, 85

[68] Billian C, Gorman PH. Upper extremity applications of functional neuromuscular stimulation. *Assistive Technology: The Official Journal of RESNA*. 1992;4(1):31–9. DOI: 10.1080/10400435.1992.10132190. 17

[69] Popovic MR, Thrasher TA, Zivanovic V, Takaki J, Hajek V. Neuroprosthesis for retraining reaching and grasping functions in severe hemiplegic patients. *Neuromodulation: Journal of the International Neuromodulation Society*. 2005 Jan;8(1):58–72. DOI: 10.1111/j.1094-7159.2005.05221.x. 18, 24

[70] Thrasher TA, Zivanovic V, McIlroy W, Popovic MR. Rehabilitation of reaching and grasping function in severe hemiplegic patients using functional electrical stimulation therapy. *Neurorehabilitation and Neural Repair*. 2008 Nov;22(6):706–14. DOI: 10.1177/1545968308317436. 18, 24

[71]   Venugopalan L, Taylor PN, Cobb JE, Swain ID. Upper limb functional electrical stimulation devices and their man-machine interfaces. *Journal of Medical Engineering & Technology*. 2015;39(8):471–9. DOI: /10.3109/03091902.2015.1102344. 18

[72]   Hebert DA, Bowen JM, Ho C, Antunes I, OReilly DJ, Bayley M. Examining a new
functional electrical stimulation therapy with people with severe upper extremity hemiparesis and chronic stroke: A feasibility study. *British Journal of Occupational Therapy*. 2017
Nov;80(11):651–9. DOI: 10.1177/0308022617719807. 19

[73]   Snoek GJ, IJzerman MJ, Hermens HJ, Maxwell D, Biering-Sorensen F. Survey of
the needs of patients with spinal cord injury: Impact and priority for improvement
in hand function in tetraplegics. *Spinal Cord*. 2004 Sep;42(9):526–32. DOI: 10.1038/
sj.sc.3101638. 19, 81

[74]   Yozbatiran N, Francisco GE. Robot-assisted therapy for the upper limb after cervical
spinal cord injury. *Physical Medicine and Rehabilitation Clinics of North America*. 2019
May;30(2):367–84. DOI: 10.1016/j.pmr.2018.12.008. 19

[75]   Ward J, Power D. Increasing the availability of nerve transfer for cervical spinal cord
injury. *Journal of Plastic, Reconstructive & Aesthetic Surgery: JPRAS*. 2016 Jul;69(7):e159.
DOI: 10.1016/j.bjps.2016.01.032. 19

[76]   Bryden A, Kilgore KL, Nemunaitis GA. Advanced assessment of the upper limb in tetraplegia: A three-tiered approach to characterizing paralysis. *Topics in Spinal Cord Injury
Rehabilitation*. 2018;24(3):206–16. DOI: 10.1310/sci2403-206. 20

[77]   O'Sullivan SB, Schmitz TJ. *Physical Rehabilitation: Assessment and Treatment*. F.A. Davis;
1994. 21

[78]   Somers MF. *Spinal Cord Injury: Functional Rehabilitation*. Appleton & Lange; 1992. 21

[79]   Colombo G, Wirz M, Dietz V. Driven gait orthosis for improvement of locomotor
training in paraplegic patients. *Spinal Cord*. 2001 May;39(5):252–5. DOI: 10.1038/
sj.sc.3101154. 21

[80]   Hesse S, Werner C, Bardeleben A. Electromechanical gait training with functional electrical stimulation: Case studies in spinal cord injury. *Spinal Cord*. 2004 Jun;42(6):346–52.
DOI: 10.1038/sj.sc.3101595. 21

[81]   Weber DJ, Stein RB, Chan KM, Loeb GE, Richmond FJR, Rolf R, et al. Functional
electrical stimulation using microstimulators to correct foot drop: A case study. *Canadian
Journal of Physiology and Pharmacology*. 2004, 82(8-9):784–92. DOI: 10.1139/y04-078. 21

[82]   Harkema SJ, Schmidt-Read M, Lorenz DJ, Edgerton VR, Behrman AL. Balance and
ambulation improvements in individuals with chronic incomplete spinal cord injury

using locomotor training-based rehabilitation. *Archives of Physical Medicine and Rehabilitation*. 2012 Sep;93(9):1508–17. DOI: 10.1016/j.apmr.2011.01.024. 21

[83] Field-Fote EC. Combined use of body weight support, functional electric stimulation, and treadmill training to improve walking ability in individuals with chronic incomplete spinal cord injury. *Archives of Physical Medicine and Rehabilitation*. 2001 Jun;82(6):818–24. DOI: 10.1053/apmr.2001.23752. 21

[84] Field-Fote EC, Lindley SD, Sherman AL. Locomotor training approaches for individuals with spinal cord injury: A preliminary report of walking-related outcomes. *Journal of Neurologic Physical Therapy: JNPT*. 2005 Sep;29(3):127–37. DOI: 10.1097/01.NPT.0000282245.31158.09. 21

[85] Andrews BJ, Baxendale RH, Barnett R. Hybrid FES orthosis incorporating closed loop control and sensory feedback. *Journal of Biomedical Engineering*. 1988 Apr;10(2):189–95. DOI: 10.1016/0141-5425(88)90099-4. 21

[86] Popovic D, Tomović R, Schwirtlich L. Hybrid assistive systemthe motor neuroprosthesis. *IEEE Transactions on Bio-Medical Engineering*. 1989 Jul;36(7):729–37. DOI: 10.1109/10.32105. 21, 46

[87] Isakov E, Douglas R, Berns P. Ambulation using the reciprocating gait orthosis and functional electrical stimulation. *Paraplegia*. 1992 Apr;30(4):239–45. DOI: 10.1038/sc.1992.62. 22

[88] Klose KJ, Jacobs PL, Broton JG, Guest RS, Needham-Shropshire BM, Lebwohl N, et al. Evaluation of a training program for persons with SCI paraplegia using the Parastep 1 ambulation system: Part 1. Ambulation performance and anthropometric measures. *Archives of Physical Medicine and Rehabilitation*. 1997 Aug;78(8):789–93. DOI: 10.1016/S0003-9993(97)90188-X. 22

[89] Thomas SL, Gorassini MA. Increases in corticospinal tract function by treadmill training after incomplete spinal cord injury. *Journal of Neurophysiology*. 2005 Oct;94(4):2844–55. DOI: 10.1152/jn.00532.2005. 23

[90] Mulcahey MJ, Betz RR, Kozin SH, Smith BT, Hutchinson D, Lutz C. Implantation of the Freehand System during initial rehabilitation using minimally invasive techniques. *Spinal Cord*. 2004 Mar;42(3):146–55. DOI: 10.1038/sj.sc.3101573. 23

[91] IJzerman MJ, Stoffers TS, in 't Groen FACG, Klatte MAP, Snoek GJ, Vorsteveld JHC, et al. The NESS handmaster orthosis: Restoration of hand function in C5 and stroke patients by means of electrical stimulation. *Journal of Rehabilitation Sciences*. 1996;9(3):86–9. 23

[92]   Prochazka A, Gauthier M, Wieler M, Kenwell Z. The bionic glove: An electrical stimulator garment that provides controlled grasp and hand opening in quadriplegia. *Archives of Physical Medicine and Rehabilitation*. 1997 Jun;78(6):608–14. DOI: 10.1016/S0003-9993(97)90426-3. 23

[93]   Rebersek S, Vodovnik L. Proportionally controlled functional electrical stimulation of hand. *Archives of Physical Medicine and Rehabilitation*. 1973 Aug;54(8):378–82. 23

[94]   Popovic DB, Popovic MB, Sinkjaer T. Neurorehabilitation of upper extremities in humans with sensory-motor impairment. *Neuromodulation: Journal of the International Neuromodulation Society*. 2002 Jan;5(1):54–66. DOI: 10.1046/j.1525-1403.2002._2009.x. 23

[95]   Kapadia N, Zivanovic V, Popovic MR. Restoring voluntary grasping function in individuals with incomplete chronic spinal cord injury: Pilot study. *Topics in Spinal Cord Injury Rehabilitation*. 2013;19(4):279–87. DOI: 10.1310/sci1904-279. 23, 24

[96]   Popovic MR, Kapadia N, Zivanovic V, Furlan JC, Craven BC, McGillivray C. Functional electrical stimulation therapy of voluntary grasping versus only conventional rehabilitation for patients with subacute incomplete tetraplegia: A randomized clinical trial. *Neurorehabilitation and Neural Repair*. 2011 Jun;25(5):433–42. DOI: 10.1177/1545968310392924. 23, 24, 85

[97]   Kapadia N, Moineau B, Popovic MR. Functional electrical stimulation therapy for retraining reaching and grasping after spinal cord injury and stroke. *Frontiers in Neuroscience*. 2020 Jul;14. DOI: 10.3389/fnins.2020.00718. 24

[98]   Kapadia NM, Bagher S, Popovic MR. Influence of different rehabilitation therapy models on patient outcomes: Hand function therapy in individuals with incomplete SCI. *The Journal of Spinal Cord Medicine*. 2014 Nov;37(6):734–43. DOI: 10.1179/2045772314Y.0000000203. 24

[99]   Popovic MR, Curt A, Keller T, Dietz V. Functional electrical stimulation for grasping and walking: Indications and limitations. *Spinal Cord*. 2001 Aug;39(8):403–12. DOI: 10.1038/sj.sc.3101191. 25

[100]   Doherty JG, Burns AS, O'Ferrall DM, Ditunno JF. Prevalence of upper motor neuron vs lower motor neuron lesions in complete lower thoracic and lumbar spinal cord injuries. *The Journal of Spinal Cord Medicine*. 2002;25(4):289–92. DOI: 10.1080/10790268.2002.11753630. 25

[101]   Stein J. Robotics in rehabilitation: Technology as destiny. *American Journal of Physical Medicine & Rehabilitation*. 2012 Nov;91(11):S199. DOI: 10.1097/PHM.0b013e31826b-cbbd. 27

[102]   Ekkelenkamp R, Veneman J, der Kooij H van. LOPES: A lower extremity powered exo-skeleton. In: *Proceedings 2007 IEEE International Conference On Robotics And Automation*. 2007. pp. 3132–3. DOI: 10.1109/ROBOT.2007.363952. 27, 28, 40

[103]   Kwakkel G, Kollen B, Lindeman E. Understanding the pattern of functional recovery after stroke: Facts and theories. *Restorative Neurology and Neuroscience*. 2004;22(3-5):281–99. 27

[104]   Hesse S, Uhlenbrock D, Werner C, Bardeleben A. A mechanized gait trainer for restoring gait in nonambulatory subjects. *Archives of Physical Medicine and Rehabilitation*. 2000 Sep;81(9):1158–61. DOI: 10.1053/apmr.2000.6280. 27, 33

[105]   Krebs HI, Hogan N, Aisen ML, Volpe BT. Robot-aided neurorehabilitation. *IEEE Transactions on Rehabilitation Engineering*. 1998 Mar;6(1):75–87. DOI: 10.1109/86.662623. 28

[106]   Jezernik S, Colombo G, Keller T, Frueh H, Morari M. Robotic orthosis lokomat: A rehabilitation and research tool. *Neuromodulation: Technology at the Neural Interface*. 2003;6(2):108–15. DOI: 10.1046/j.1525-1403.2003.03017.x. 28, 38, 39

[107]   Yue Z, Zhang X, Wang J. Hand rehabilitation robotics on poststroke motor recovery. Vol. 2017, *Behavioral Neurology*. https://www.hindawi.com/journals/bn/2017/3908135/; Hindawi; 2017. p. e3908135. DOI: 10.1155/2017/3908135. 28

[108]   Weber LM, Stein J. The use of robots in stroke rehabilitation: A narrative review. *NeuroRehabilitation*. 2018 Jan;43(1):99–110. DOI: 10.3233/NRE-172408. 28, 49

[109]   Maciejasz P, Eschweiler J, Gerlach-Hahn K, Jansen-Troy A, Leonhardt S. A survey on robotic devices for upper limb rehabilitation. *Journal of NeuroEngineering and Rehabilitation*. 2014 Jan;11(1):3. DOI: 10.1186/1743-0003-11-3. 28

[110]   Lee SH, Park G, Cho DY, Kim HY, Lee J-Y, Kim S, et al. Comparisons between end-effector and exoskeleton rehabilitation robots regarding upper extremity function among chronic stroke patients with moderate-to-severe upper limb impairment. *Scientific Reports*. 2020 Feb;10(1):1806. DOI:10.1038/s41598-020-58630-2. 28

[111]   Hesse S, Schmidt H, Werner C, Bardeleben A. Upper and lower extremity robotic devices for rehabilitation and for studying motor control. *Current Opinion in Neurology*. 2003 Dec;16(6):705–10. DOI: 10.1097/00019052-200312000-00010. 28, 29, 31, 33, 38

[112]   Krebs HI, Volpe BT, Williams D, Celestino J, Charles SK, Lynch D, et al. Robot-aided neurorehabilitation: A robot for wrist rehabilitation. *IEEE Transactions on Neural Systems and Rehabilitation Engineering: A Publication of the IEEE Engineering in Medicine and Biology Society*. 2007 Sep;15(3):327–35. DOI: 10.1109/TNSRE.2007.903899. 29

[113]   Lum PS, Burgar CG, Shor PC. Evidence for improved muscle activation patterns after retraining of reaching movements with the MIME robotic system in subjects with post-stroke hemiparesis. *IEEE Transactions on Neural Systems and Rehabilitation Engineering.* 2004 Jun;12(2):186–94. DOI: 10.1109/TNSRE.2004.827225. 29

[114]   Lum P, Reinkensmeyer D, Mahoney R, Rymer WZ, Burgar C. Robotic devices for movement therapy after stroke: Current status and challenges to clinical acceptance. *Topics in Stroke Rehabilitation.* 2002;8(4):40–53. DOI: 10.1310/9KFM-KF81-P9A4-5WW0. 30

[115]   Reinkensmeyer DJ, Schmit BD, Rymer WZ. Assessment of active and passive restraint during guided reaching after chronic brain injury. *Annals of Biomedical Engineering.* 1999; 27(6):805–14. DOI: 10.1114/1.233. 30

[116]   Reinkensmeyer DJ, Dewald JP, Rymer WZ. Guidance-based quantification of arm impairment following brain injury: A pilot study. *IEEE Transactions on Rehabilitation Engineering: A Publication of the IEEE Engineering in Medicine and Biology Society.* 1999 Mar;7(1):1–11. DOI: 10.1109/86.750543. 30

[117]   Hesse S, Schmidt H, Werner C. Machines to support motor rehabilitation after stroke: 10 years of experience in Berlin. *Journal of Rehabilitation Research and Development.* 2006; 43(5):671–8. DOI: 10.1682/JRRD.2005.02.0052. 31, 36

[118]   Loureiro R, Amirabdollahian F, Topping M, Driessen B, Harwin W. Upper limb robot mediated stroke therapy GENTLE/s Approach. *Autonomous Robots.* 2003 Jul;15(1):35–51. DOI: 10.1023/A:1024436732030. 32

[119]   Masiero S, Celia A, Armani M, Rosati G. A novel robot device in rehabilitation of post-stroke hemiplegic upper limbs. *Aging Clinical and Experimental Research.* 2006 Dec;18(6):531–5. DOI: 10.1007/BF03324854. 32, 33

[120]   Rosati G, Gallina P, Masiero S. Design, Implementation and clinical tests of a wire-based robot for neurorehabilitation. *IEEE Transactions on Neural Systems and Rehabilitation Engineering.* 2007 Dec;15(4):560–9. DOI: 10.1109/TNSRE.2007.908560. 32, 33

[121]   Hesse S, Werner C, Uhlenbrock D, von Frankenberg S, Bardeleben A, Brandl-Hesse B. An electromechanical gait trainer for restoration of gait in hemiparetic stroke patients: Preliminary results. *Neurorehabilitation and Neural Repair.* 2001;15(1):39–50. DOI: 10.1177/154596830101500106. 33, 34

[122]   Schmidt H, Krüger J, Hesse S. HapticWalker Haptic foot device for gait rehabilitation. In: *Human Haptic Perception: Basics and Applications.* Birkhäuser Basel; 2008. pp. 501–11. DOI: 10.1007/978-3-7643-7612-3_42. 35

[123] Lo HS, Xie SQ. Exoskeleton robots for upper-limb rehabilitation: State of the art and future prospects. *Medical Engineering & Physics*. 2012 Apr;34(3):261–8. DOI: 10.1016/j.medengphy.2011.10.004. 36

[124] Pignolo L. Robotics in neuro-rehabilitation. *Journal of Rehabilitation Medicine*. 2009 Nov;41(12):955–60. DOI: 10.2340/16501977-0434. 36

[125] Nef T, Mihelj M, Riener R. ARMin: A robot for patient-cooperative arm therapy. *Medical & Biological Engineering & Computing*. 2007 Sep;45(9):887–900. DOI: 10.1007/s11517-007-0226-6. 36, 37

[126] Guidali M, Duschau-Wicke A, Broggi S, Klamroth-Marganska V, Nef T, Riener R. A robotic system to train activities of daily living in a virtual environment. *Medical & Biological Engineering & Computing*. 2011 Jul;49(10):1213. DOI: 10.1007/s11517-011-0809-0. 37

[127] Zariffa J, Kapadia N, Kramer JLK, Taylor P, Alizadeh-Meghrazi M, Zivanovic V, et al. Effect of a robotic rehabilitation device on upper limb function in a sub-acute cervical spinal cord injury population. *IEEE International Conference on Rehabilitation Robotics*: [proceedings]. 2011:5975400. DOI: 10.1109/ICORR.2011.5975400. 37, 38

[128] Vukobratovic M, Hristic D, Stojiljkovic Z. Development of active anthropomorphic exoskeletons. *Medical & Biological Engineering*. 1974 Jan;12(1):66–80. DOI: 10.1007/BF02629836. 38

[129] Veneman JF, Kruidhof R, Hekman EEG, Ekkelenkamp R, Van Asseldonk EHF, van der Kooij H. Design and evaluation of the LOPES exoskeleton robot for interactive gait rehabilitation. *IEEE Transactions on Neural Systems and Rehabilitation Engineering: A Publication of the IEEE Engineering in Medicine and Biology Society*. 2007 Sep;15(3):379–86. DOI: 10.1109/TNSRE.2007.903919. 39

[130] Calabrò RS, Cacciola A, Bertè F, Manuli A, Leo A, Bramanti A, et al. Robotic gait rehabilitation and substitution devices in neurological disorders: Where are we now? *Neurological Sciences: Official Journal of the Italian Neurological Society and of the Italian Society of Clinical Neurophysiology*. 2016 Apr;37(4):503–14. DOI: 10.1007/s10072-016-2474-4. 40

[131] Escalona MJ, Bourbonnais D, Goyette M, Duclos C, Gagnon DH. Wearable exoskeleton control modes selected during overground walking affect muscle synergies in adults with a chronic incomplete spinal cord injury. *Spinal Cord Series and Cases*. 2020 Apr;6(1):26. DOI: 10.1038/s41394-020-0269-6. 41

[132]   Kawamoto H, Sankai Y. Comfortable power assist control method for walking aid by HAL-3. In: *IEEE International Conference on Systems, Man and Cybernetics*. 2002. p. 6 vol.4. 41

[133]   Veerbeek JM, Langbroek-Amersfoort AC, van Wegen EEH, Meskers CGM, Kwakkel G. Effects of robot-assisted therapy for the upper limb after stroke. *Neurorehabilitation and Neural Repair*. 2017 Feb;31(2):107–21. DOI: 10.1177/1545968316666957. 41

[134]   Chen Z, Wang C, Fan W, Gu M, Yasin G, Xiao S, et al. Robot-assisted arm training versus therapist-mediated training after stroke: A systematic review and meta-analysis. Vol. 2020, *Journal of Healthcare Engineering*. https://www.hindawi.com/journals/jhe/2020/8810867/; Hindawi; 2020. p. e8810867. DOI: 10.1155/2020/8810867. 42

[135]   Mehrholz J, Pollock A, Pohl M, Kugler J, Elsner B. Systematic review with network meta-analysis of randomized controlled trials of robotic-assisted arm training for improving activities of daily living and upper limb function after stroke. *Journal of NeuroEngineering and Rehabilitation*. 2020 Jun;17(1):83. DOI: 10.1186/s12984-020-00715-0. 42

[136]   Wu J, Cheng H, Zhang J, Yang S, Cai S. Robot-assisted therapy for upper extremity motor impairment after stroke: A systematic review and meta-analysis. *Physical Therapy*. 2021; Apr 4;101(4):pzab010. DOI: 10.1093/ptj/pzab010. DOI: 10.1093/ptj/pzab010. 42

[137]   Mehrholz J, Thomas S, Werner C, Kugler J, Pohl M, Elsner B. Electromechanical-assisted training for walking after stroke. *The Cochrane Database of Systematic Reviews*. 2017 May;5:CD006185. DOI: 10.1002/14651858.CD006185.pub4. 42

[138]   Zheng Q-X, Ge L, Wang CC, Ma Q-S, Liao Y-T, Huang P-P, et al. Robot-assisted therapy for balance function rehabilitation after stroke: A systematic review and meta-analysis. *International Journal of Nursing Studies*. 2019;95(Complete):7–18. DOI: 10.1016/j.ijnurstu.2019.03.015. 43

[139]   Singh H, Unger J, Zariffa J, Pakosh M, Jaglal S, Craven BC, et al. Robot-assisted upper extremity rehabilitation for cervical spinal cord injuries: A systematic scoping review. *Disability and Rehabilitation Assistive Technology*. 2018 Oct;13(7):704–15. DOI: 10.1080/17483107.2018.1425747. 43

[140]   Cheung EYY, Ng TKW, Yu KKK, Kwan RLC, Cheing GLY. Robot-assisted training for people with spinal cord injury: A meta-analysis. *Archives of Physical Medicine and Rehabilitation*. 2017 Nov;98(11):2320–2331.e12. DOI: 10.1016/j.apmr.2017.05.015. 43, 44

[141]   Fang C-Y, Tsai J-L, Li G-S, Lien AS-Y, Chang Y-J. Effects of robot-assisted gait training in individuals with spinal cord injury: A meta-analysis. *BioMed Research International*. 2020:2102785. DOI: 10.1155/2020/2102785. 44

[142]   del-Ama AJ, Koutsou AD, Moreno JC, de-los-Reyes A, Gil-Agudo A, Pons JL. Review of hybrid exoskeletons to restore gait following spinal cord injury. *Journal of Rehabilitation Research and Development*. 2012;49(4):497–514. DOI: 10.1682/JRRD.2011.03.0043. 44

[143]   To CS, Kobetic R, Schnellenberger JR, Audu ML, Triolo RJ. Design of a variable constraint hip mechanism for a hybrid neuroprosthesis to restore gait after spinal cord injury. *IEEE/ASME Transactions on Mechatronics*. 2008 Apr;13(2):197–205. DOI: 10.1109/TMECH.2008.918551. 44

[144]   To CS, Kobetic R, Triolo RJ. Hybrid orthosis system with a variable hip coupling mechanism. In: *2006 International Conference of the IEEE Engineering in Medicine and Biology Society*. 2006. pp. 2928–31. DOI: 10.1109/IEMBS.2006.259631. 44

[145]   Durfee WK, Goldfarb M. Design of a controlled-brake orthosis for regulating FES-aided gait. In: *1992 14th Annual International Conference of the IEEE Engineering in Medicine and Biology Society*. 1992. pp. 1337–8. 45

[146]   Farris RJ, Quintero HA, Withrow TJ, Goldfarb M. Design of a joint-coupled orthosis for FES-aided gait. In: *2009 IEEE International Conference on Rehabilitation Robotics*. 2009. pp. 246–52. DOI: 10.1109/ICORR.2009.5209623. 45

[147]   Durfee WK. Design and simulation of a pneumatic, stored-energy, hybrid orthosis for gait restoration. *Journal of Biomechanical Engineering*. 2005 Jul;127(6):1014. DOI: 10.1115/1.2050652. 45

[148]   Kangude A, Burgstahler B, Kakastys J, Durfee W. Single channel hybrid FES gait system using an energy storing orthosis: Preliminary design. In: *2009 Annual International Conference of the IEEE Engineering in Medicine and Biology Society*. 2009. pp. 6798–801. DOI: 10.1109/IEMBS.2009.5333976. 45

[149]   Stauffer Y, Allemand Y, Bouri M, Fournier J, Clavel R, Metrailler P, et al. The WalkTrainer—A new generation of walking reeducation device combining orthoses and muscle stimulation. *IEEE Transactions on Neural Systems and Rehabilitation Engineering*. 2009 Feb;17(1):38–45. DOI: 10.1109/TNSRE.2008.2008288. 45, 46

[150]   Pedrocchi A, Ferrante S, Ambrosini E, Gandolla M, Casellato C, Schauer T, et al. MUNDUS project: MUltimodal neuroprosthesis for daily upper limb support. *Journal of NeuroEngineering and Rehabilitation*. 2013 Jul;10:66. DOI: 10.1186/1743-0003-10-66. 46, 47

[151]   Ambrosini E, Zajc J, Ferrante S, Ferrigno G, Gasperina SD, Bulgheroni M, et al. A hybrid robotic system for arm training of stroke survivors: Concept and first evalua-

tion. *IEEE Transactions on Bio-Medical Engineering*. 2019 Dec;66(12):3290–300. DOI: 10.1109/TBME.2019.2900525. 46, 47

[152]   Scott S, Yu T, White KT, Van Harlinger W, Gonzalez Y, Llanos I, et al. A robotic hand device safety study for people with cervical spinal cord injury. *Federal Practitioner: For The Health Care Professionals of the VA, DoD, and PHS*. 2018 Apr;35(Suppl 3):S21–5. 48

[153]   Stein J. Robotics in rehabilitation: Technology as destiny. *American Journal of Physical Medicine & Rehabilitation*. 2012 Nov;91(11 Suppl 3):S199–203. DOI: 10.1097/PHM.0b013e31826bcbbd. 48

[154]   Wolpaw JR, Birbaumer N, Heetderks WJ, McFarland DJ, Peckham PH, Schalk G, et al. Brain–computer interface technology: A review of the first international meeting. *IEEE Transactions on Rehabilitation Engineering*. 2000 Jun;8(2):164–73. DOI: 10.1109/TRE.2000.847807. 51

[155]   Vidal JJ. Real-time detection of brain events in EEG. *Proceedings of the IEEE*. 1977;65(5):633–41. DOI: 10.1109/PROC.1977.10542. 51, 56

[156]   Vidal J. Toward direct brain–computer communication. *Annual Review of Biophysics and Bioengineering*. 1973;2(1):157–80. DOI: 10.1146/annurev.bb.02.060173.001105. 51

[157]   Hochberg LR, Serruya MD, Friehs GM, Mukand JA, Saleh M, Caplan AH, et al. Neuronal ensemble control of prosthetic devices by a human with tetraplegia. *Nature*. 2006 Jul;442(7099):164–71. DOI: 10.1038/nature04970. 53, 62

[158]   Ajiboye AB, Simeral JD, Donoghue JP, Hochberg LR, Kirsch RF. Prediction of imagined single-joint movements in a person with high-level tetraplegia. *IEEE Transactions on Bio-Medical Engineering*. 2012;59(10):2755–65. DOI: 10.1109/TBME.2012.2209882. 53

[159]   Chadwick EK, Blana D, Simeral JD, Lambrecht J, Kim SP, Cornwell AS, et al. Continuous neuronal ensemble control of simulated arm reaching by a human with tetraplegia. *Journal of Neural Engineering*. 2011 May;8(3):034003. DOI: 10.1088/1741-2560/8/3/034003. 53

[160]   Collinger JL, Wodlinger B, Downey JE, Wang W, Tyler-Kabara EC, Weber DJ, et al. High-performance neuroprosthetic control by an individual with tetraplegia. *The Lancet*. 2013 Feb;381(9866):557–64. DOI: 10.1016/S0140-6736(12)61816-9. 53, 61, 62

[161]   Ajiboye AB, Willett FR, Young DR, Memberg WD, Murphy BA, Miller JP, et al. Restoration of reaching and grasping movements through brain-controlled muscle stimulation in a person with tetraplegia: A proof-of-concept demonstration. *The Lancet*. 2017 May;389(10081):1821–30. DOI: 10.1016/S0140-6736(17)30601-3. 53, 63, 65

[162] Toro C, Cox C, Friehs G, Ojakangas C, Maxwell R, Gates JR, et al. 8-12 Hz rhythmic oscillations in human motor cortex during two-dimensional arm movements: Evidence for representation of kinematic parameters. *Electroencephalography and Clinical Neurophysiology*. 1994 Oct;93(5):390–403. DOI: 10.1016/0168-5597(94)90127-9. 53

[163] Rickert J, Oliveira SC de, Vaadia E, Aertsen A, Rotter S, Mehring C. Encoding of movement direction in different frequency ranges of motor cortical local field potentials. *The Journal of Neuroscience: The Official Journal of the Society for Neuroscience*. 2005 Sep;25(39):8815–24. DOI: 10.1523/JNEUROSCI.0816-05.2005. 53

[164] Mehring C, Nawrot M, de Oliveira S, Vaadia E, Schulze-Bonhage A, Aertsen A, et al. Comparing information about arm movement direction in single channels of local and epicortical field potentials from monkey and human motor cortex. *Journal of Physiology-Paris*. 2004;98(4-6):498–506. DOI: 10.1016/j.jphysparis.2005.09.016. 53

[165] Milekovic T, Fischer J, Pistohl T, et al. An online brain-Machine interface using decoding of movement direction from the human electrocorticogram. *Journal of Neural Engineering*. 2012; 9(4):046003. DOI: 10.1088/1741-2560/9/4/046003. 53, 61

[166] Anderson NR, Blakely T, Schalk G, Leuthardt EC, Moran DW. Electrocorticographic (ECoG) correlates of human arm movements. *Experimental Brain Research*. 2012 Nov;223(1):1–10. DOI: 10.1007/s00221-012-3226-1. 53

[167] Kubánek J, Miller KJ, Ojemann JG, Wolpaw JR, Schalk G. Decoding flexion of individual fingers using electrocorticographic signals in humans. *Journal of Neural Engineering*. 2009 Dec;6(6):066001. DOI: 10.1088/1741-2560/6/6/066001. 53

[168] Leuthardt EC, Schalk G, Wolpaw JR, Ojemann JG, Moran DW. A brain–computer interface using electrocorticographic signals in humans. *Journal of Neural Engineering*. 2004 Jun;1(2):63–71. DOI: 10.1088/1741-2560/1/2/001. 53, 61

[169] Marquez-Chin C, Sanin E, Silva J, Popovic M. Real-time two-dimensional asynchronous control of a remote-controlled car using a single electroencephalographic electrode. *Topics in Spinal Cord Injury Rehabilitation*. 2009 May;14(4):62–8. DOI: 10.1310/sci1404-62. 53

[170] Marquez-Chin C, Popovic MR, Sanin E, Chen R, Lozano AM. Real-time two-dimensional asynchronous control of a computer cursor with a single subdural electrode. *The Journal of Spinal Cord Medicine*. 2012 Sep;35(5):382–91. DOI: 10.1179/2045772312Y.0000000043. 53, 61, 90

[171] Márquez-Chin C, Popovic MR, Cameron T, Lozano AM, Chen R. Control of a neuroprosthesis for grasping using off-line classification of electrocorticographic signals: Case study. *Spinal Cord*. 2009 Apr;47(11):802–8. DOI: 10.1038/sc.2009.41. 53, 64

[172]   Spüler M, Walter A, Ramos-Murguialday A, Naros G, Birbaumer N, Gharabaghi A, et al. Decoding of motor intentions from epidural ECoG recordings in severely paralyzed chronic stroke patients. *Journal of Neural Engineering.* 2014 Oct;11(6):066008. DOI: 10.1088/1741-2560/11/6/066008. 54

[173]   Slutzky MW. Brain-machine interfaces: Powerful tools for clinical treatment and neuroscientific investigations. *The Neuroscientist.* 2018 May;25(2):139–54. DOI: 10.1177/1073858418775355. 54

[174]   Glover GH. Overview of functional magnetic resonance imaging. *Neurosurgery Clinics of North America.* 2011 Apr;22(2):133–9. DOI: 10.1016/j.nec.2010.11.001. 54

[175]   Almajidy RK, Mankodiya K, Abtahi M, Hofmann UG. A Newcomer's Guide to Functional Near Infrared Spectroscopy Experiments. *IEEE Reviews in Biomedical Engineering.* 2020;13:292–308. DOI: 10.1109/RBME.2019.2944351. 54

[176]   Yang M, Yang Z, Yuan T, Feng W, Wang P. A systemic review of functional near-infrared spectroscopy for stroke: Current application and future directions. *Frontiers in Neurology.* 2019;10. DOI: 10.3389/fneur.2019.00058. 54

[177]   Power SD, Kushki A, Chau T. Automatic single-trial discrimination of mental arithmetic, mental singing and the no-control state from prefrontal activity: Toward a three-state NIRS-BCI. *BMC Research Notes.* 2012; DOI: 10.1186/1756-0500-5-141. 54, 56

[178]   Caria A, Weber C, Brötz D, Ramos A, Ticini LF, Gharabaghi A, et al. Chronic stroke recovery after combined BCI training and physiotherapy: A case report. *Psychophysiology.* 2010 Aug;48(4):578–82. DOI: 10.1111/j.1469-8986.2010.01117.x. 55

[179]   Battapady H, Lin P, Holroyd T, Hallett M, Chen X, Fei D-Y, et al. Spatial detection of multiple movement intentions from SAM-filtered single-trial MEG signals. *Clinical Neurophysiology.* 2009 Nov;120(11):1978–87. DOI: 10.1016/j.clinph.2009.08.017. 55

[180]   Berger H. Über das elektrenkephalogramm des menschen. *Archiv für Psychiatrie und Nervenkrankheiten.* 1929 Dec;87(1):527–70. DOI: 10.1007/BF01797193. 55

[181]   Cooper R, Binnie CD, Billings R. *Techniques in Clinical Neurophysiology: A Practical Manual.* Elsevier Churchill Livingstone; 2005. 55

[182]   Mason S, Bashashati A, Fatourechi M, Navarro K, Birch G. A comprehensive survey of brain interface technology designs. *Annals of Biomedical Engineering.* 2007;35(2):137–69. DOI: 10.1007/s10439-006-9170-0. 55

[183]   Leeb R, Friedman D, Müller-Putz GR, Scherer R, Slater M, Pfurtscheller G. Self-paced (asynchronous) BCI control of a wheelchair in virtual environments: A case

study with a tetraplegic. *Computational Intelligence and Neuroscience*. 2007; (2):1–8. DOI: 10.1155/2007/79642. 56, 62, 90

[184] Pfurtscheller G, Lopes da Silva FH. Event-related EEG/MEG synchronization and desynchronization: Basic principles. *Clinical Neurophysiology: Official Journal of the International Federation of Clinical Neurophysiology*. 1999;110(11):1842–57. DOI: 10.1016/S1388-2457(99)00141-8. 57, 59

[185] Graimann B, Huggins JE, Levine SP, Pfurtscheller G. Visualization of significant ERD/ERS patterns in multichannel EEG and ECoG data. *Clinical Neurophysiology*. 2002 Jan;113(1):43–7. DOI: 10.1016/S1388-2457(01)00697-6. 58, 59

[186] Bai O, Rathi V, Lin P, Huang D, Battapady H, Fei D-Y, et al. Prediction of human voluntary movement before it occurs. *Clinical Neurophysiology: Official Journal of the International Federation of Clinical Neurophysiology*. 2011 Feb;122(2):364–72. DOI: 10.1016/j.clinph.2010.07.010. 60

[187] Jochumsen M, Niazi IK, Dremstrup K, Kamavuako EN. Detecting and classifying three different hand movement types through electroencephalography recordings for neurorehabilitation. *Medical & Biological Engineering & Computing*. 2015 Dec;54(10):1491–501. DOI: 10.1007/s11517-015-1421-5. 60

[188] Xu R, Jiang N, Mrachacz-Kersting N, Lin C, Asín Prieto G, Moreno JC, et al. A closed-loop brain–computer interface triggering an active ankle-foot orthosis for inducing cortical neural plasticity. *IEEE Transactions on Bio-Medical Engineering*. 2014 Jul;61(7):2092–101. DOI: 10.1109/TBME.2014.2313867. 60

[189] Niazi IK, Jiang N, Tiberghien O, Nielsen JF, Dremstrup K, Farina D. Detection of movement intention from single-trial movement-related cortical potentials. *Journal of Neural Engineering*. 2011 Dec;8(6):066009. DOI: 10.1088/1741-2560/8/6/066009. 60

[190] Bai O, Lin P, Vorbach S, Floeter MK, Hattori N, Hallett M. A high performance sensorimotor beta rhythm-based brain–computer interface associated with human natural motor behavior. *Journal of Neural Engineering*. 2008 Mar;5(1):24–35. DOI: 10.1088/1741-2560/5/1/003. 60

[191] Xu R, Jiang N, Vuckovic A, Hasan M, Mrachacz-Kersting N, Allan D, et al. Movement-related cortical potentials in paraplegic patients: Abnormal patterns and considerations for BCI-rehabilitation. *Frontiers in Neuroengineering*. 2014;7:35. DOI: 10.3389/fneng.2014.00035. 60

[192] Mrachacz-Kersting N, Jiang N, Stevenson AJT, Niazi IK, Kostic V, Pavlovic A, et al. Efficient neuroplasticity induction in chronic stroke patients by an associative brain–com-

puter interface. *Journal of Neurophysiology*. 2016 Mar;115(3):1410–21. DOI: 10.1152/jn.00918.2015. 60, 79, 80, 81, 104

[193]   Donchin E, Spencer KM, Wijesinghe R. The mental prosthesis: Assessing the speed of a P300-based brain–computer interface. *IEEE Transactions on Rehabilitation Engineering*. 2000 Jun;8(2):174–9. DOI: 10.1109/86.847808. 60

[194]   Lenhardt A, Ritter H. *An Augmented-Reality Based Brain–computer Interface For Robot Control*. In Berlin, Heidelberg: Springer Berlin Heidelberg; 2010. pp. 58–65. DOI: 10.1007/978-3-642-17534-3_8. 61

[195]   Kouji Takano NHKK. Towards intelligent environments: An augmented RealityBrain-Machine interface operated with a see-through head-mount display. *Frontiers in Neuroscience*. 2011;5. DOI: 10.3389/fnins.2011.00060. 61

[196]   Brouwer A-M, van Erp JBF. A tactile p300 brain–computer interface. *Frontiers in Neuroscience*. 2010;4:19. DOI: 10.3389/fnins.2010.00019. 61

[197]   Wolpaw JR, McFarland DJ. Control of a two-dimensional movement signal by a noninvasive brain–computer interface in humans. *Proceedings of the National Academy of Sciences*. 2004 Dec;101(51):17849–54. DOI: 10.1073/pnas.0403504101. 61

[198]   Krusienski DJ, Schalk G, McFarland DJ, Wolpaw JR. A mu-rhythm matched filter for continuous control of a brain–computer interface. *IEEE Transactions on Bio-Medical Engineering*. 2007 Feb;54(2):273–80. DOI: 10.1109/TBME.2006.886661. 61

[199]   McFarland DJ, Krusienski DJ, Sarnacki WA, Wolpaw JR. Emulation of computer mouse control with a noninvasive brainComputer interface. *Journal of Neural Engineering*. 2008 Mar;5(2):101–10. DOI: 10.1088/1741-2560/5/2/001. 61

[200]   Huang D, Lin P, Fei D-Y, Chen X, Bai O. Decoding human motor activity from EEG single trials for a discrete two-dimensional cursor control. *Journal of Neural Engineering*. 2009;6(4):046005. DOI: 10.1088/1741-2560/6/4/046005. 61

[201]   Schalk G, Miller KJ, Anderson NR, Wilson JA, Smyth MD, Ojemann JG, et al. Two-dimensional movement control using electrocorticographic signals in humans. *Journal of Neural Engineering*. 2008 Mar;5(1):75–84. DOI: 10.1088/1741-2560/5/1/008. 61

[202]   Marquez-Chin C, Popovic MR, Sanin E, Chen R, Lozano AM. Real-time two-dimensional asynchronous control of a computer cursor with a single subdural electrode. *The Journal of Spinal Cord Medicine*. 2012 Sep;35(5):382–91. DOI: 10.1179/2045772312Y.0000000043.

[203]   Kim S-P, Simeral JD, Hochberg LR, Donoghue JP, Friehs GM, Black MJ. Point-and-click cursor control with an intracortical neural interface system by humans with tetraple-

gia. *IEEE Transactions on Neural Systems and Rehabilitation Engineering.* 2011;19(2):193–203. DOI: 10.1109/TNSRE.2011.2107750. 61

[204]   LaFleur K, Cassady K, Doud A, Shades K, Rogin E, He B. Quadcopter control in three-dimensional space using a noninvasive motor imagery-based brain–computer interface. *Journal of Neural Engineering.* 2013 Aug;10(4):046003. DOI: 10.1088/1741-2560/10/4/046003. 61

[205]   McFarland DJ, Sarnacki WA, Wolpaw JR. Electroencephalographic (EEG) control of three-dimensional movement. *Journal of Neural Engineering.* 2010 May;7(3):036007. DOI: 10.1088/1741-2560/7/3/036007. 61

[206]   Wang W, Collinger JL, Degenhart AD, Tyler-Kabara EC, Schwartz AB, Moran DW, et al. An electrocorticographic brain interface in an individual with tetraplegia. *PloS One.* 2013;8(2):e55344. DOI: 10.1371/journal.pone.0055344. 61

[207]   Bell CJ, Shenoy P, Chalodhorn R, Rao RPN. Control of a humanoid robot by a noninvasive brainComputer interface in humans. *Journal of Neural Engineering.* 2008 May;5(2):214–20. DOI: 10.1088/1741-2560/5/2/012. 61

[208]   Collinger JL, Wodlinger B, Downey JE, Wang W, Tyler-Kabara EC, Weber DJ, et al. High-performance neuroprosthetic control by an individual with tetraplegia. *The Lancet.* 2013 Feb;381(9866):557–64. DOI: 10.1016/S0140-6736(12)61816-9.

[209]   Hochberg LR, Bacher D, Jarosiewicz B, Masse NY, Simeral JD, Vogel J, et al. Reach and grasp by people with tetraplegia using a neurally controlled robotic arm. *Nature.* 2012 May;485(7398):372–5. DOI: 10.1038/nature11076. 61

[210]   Onose G, Grozea C, Anghelescu A, Daia C, Sinescu CJ, Ciurea AV, et al. On the feasibility of using motor imagery EEG-based brain|[Ndash]|computer interface in chronic tetraplegics for assistive robotic arm control: A clinical test and long-term post-trial follow-up. *Spinal Cord.* 2012 Aug;50(8):599–608. DOI: 10.1038/sc.2012.14. 61

[211]   Rao RPN, Stocco A, Bryan M, Sarma D, Youngquist TM, Wu J, et al. A direct brain-to-brain interface in humans. *PloS One.* 2014 Nov;9(11):e111332. DOI: 10.1371/journal.pone.0111332. 61

[212]   Pfurtscheller G, Guger C, Müller G, Krausz G, Neuper C. Brain oscillations control hand orthosis in a tetraplegic. *Neuroscience Letters.* 2000 Oct;292(3):211–4. DOI: 10.1016/S0304-3940(00)01471-3. 61, 64, 90

[213]   Wang PT, King CE, Chui LA, Do AH, Nenadic Z. Self-paced brainComputer interface control of ambulation in a virtual reality environment. *Journal of Neural Engineering.* 2012 Oct;9(5):056016. DOI: 10.1088/1741-2560/9/5/056016. 63

[214]   King CE, Wang PT, Chui LA, Do AH, Nenadic Z. Operation of a brain–computer interface walking simulator for individuals with spinal cord injury. *Journal of NeuroEngineering and Rehabilitation.* 2013;10(1):77. DOI: 10.1186/1743-0003-10-77. 63

[215]   Do AH, Wang PT, King CE, Chun SN, Nenadic Z. Brain–computer interface controlled robotic gait orthosis. *Journal of NeuroEngineering and Rehabilitation.* 2013;10:111. DOI: DOI: 10.1186/1743-0003-10-111. 63

[216]   López-Larraz E, Trincado-Alonso F, Rajasekaran V, Pérez-Nombela S, del-Ama AJ, Aranda J, et al. Control of an ambulatory exoskeleton with a BrainMachine interface for spinal cord injury gait rehabilitation. *Frontiers in Neuroscience.* 2016 Aug;10(122):e103764. DOI: 10.3389/fnins.2016.00359. 63

[217]   Pfurtscheller G, Müller GR, Pfurtscheller J, Gerner HJ, Rupp R. "Thought" control of functional electrical stimulation to restore hand grasp in a patient with tetraplegia. *Neuroscience Letters.* 2003 Nov;351(1):33–6. DOI: 10.1016/S0304-3940(03)00947-9. 64

[218]   Müller-Putz GR, Scherer R, Pfurtscheller G, Rupp R. EEG-based neuroprosthesis control: A step towards clinical practice. *Neuroscience Letters.* 2005;382(1-2):169–74. DOI: 10.1016/j.neulet.2005.03.021. 64

[219]   Rohm M, Schneiders M, Müller C, Kreilinger A, Kaiser V, Müller-Putz GR, et al. Hybrid brainComputer interfaces and hybrid neuroprostheses for restoration of upper limb functions in individuals with high-level spinal cord injury. *Artificial Intelligence in Medicine.* 2013 Oct;59(2):133–42. DOI: 10.1016/j.artmed.2013.07.004. 65

[220]   Friedenberg DA, Schwemmer MA, Landgraf AJ, Annetta NV, Bockbrader MA, Bouton CE, et al. Neuroprosthetic-enabled control of graded arm muscle contraction in a paralyzed human. *Scientific Reports.* 2017 Aug;7(1):8386. DOI: 10.1038/s41598-017-08120-9. 65

[221]   Dimyan MA, Cohen LG. Neuroplasticity in the context of motor rehabilitation after stroke. *Nature Reviews Neurology.* 2011 Feb;7(2):76–85. DOI: 10.1038/nrneurol.2010.200. 67, 102

[222]   Dobkin BH. Brain–computer interface technology as a tool to augment plasticity and outcomes for neurological rehabilitation. *The Journal of Physiology.* 2007 Mar;579(3):637–42. DOI: 10.1113/jphysiol.2006.123067. 67

[223]   Butler AJ, Page SJ. Mental practice with motor imagery: Evidence for motor recovery and cortical reorganization after stroke. *Archives of Physical Medicine and Rehabilitation.* 2006 Dec;87(12):2–11. DOI: 10.1016/j.apmr.2006.08.326. 68

[224] Page Stephen J., Peters Heather. Mental practice. *Stroke.* 2014 Nov;45(11):3454–60. DOI: 10.1161/STROKEAHA.114.004313. 68

[225] Upper Extremity Interventions | EBRSR - Evidence-Based Review of Stroke Rehabilitation. http://www.ebrsr.com/evidence-review/10-upper-extremity-interventions. 68

[226] Ang KK, Chua KSG, Phua KS, Wang C, Chin ZY, Kuah CWK, et al. A randomized controlled trial of EEG-Based motor imagery brain–computer interface robotic rehabilitation for stroke. *Clinical EEG and Neuroscience.* 2014 Sep;46(4):310–20. DOI: 10.1177/1550059414522229. 70, 71, 104

[227] Ang KK, Guan C, Sui Geok Chua K, Ang BT, Kuah C, Wang C, et al. A clinical study of motor imagery-based brain–computer interface for upper limb robotic rehabilitation. In: 2009 Annual International Conference of the IEEE Engineering in Medicine and Biology Society. Institute for Infocomm Research, Agency for Science, Technology and Research, *IEEE*; 2009. pp. 5981–4. 70

[228] Daly JJ, Cheng R, Rogers J, Litinas K, Hrovat K, Dohring M. Feasibility of a new application of noninvasive brain computer interface (BCI): A case study of training for recovery of volitional motor control after stroke. *Journal of Neurologic Physical Therapy.* 2009 Dec;33(4):203–11. DOI: 10.1097/NPT.0b013e3181c1fc0b. 71, 76, 77

[229] Mihara M, Hattori N, Hatakenaka M, Yagura H, Kawano T, Hino T, et al. Near-infrared spectroscopy-mediated neurofeedback enhances efficacy of motor imagery-based training in poststroke victims: A pilot study. *Stroke.* 2013 Apr;44(4):1091–8. DOI: 10.1161/STROKEAHA.111.674507. 71, 76, 77

[230] Li M, Liu Y, Wu Y, Liu S, Jia J, Zhang L. Neurophysiological substrates of stroke patients with motor imagery-based brain–computer interface training. *International Journal of Neuroscience.* 2013 Oct. DOI: 10.3109/00207454.2013.850082. 72, 76, 77

[231] Ono T, Shindo K, Kawashima K, Ota N, Ito M, Ota T, et al. Brain–computer interface with somatosensory feedback improves functional recovery from severe hemiplegia due to chronic stroke. *Frontiers in Neuroengineering.* 2014;7:19. DOI: 10.3389/fneng.2014.00019. 73, 76, 77

[232] Tsuji T, Liu M, Sonoda S, Domen K, Chino N. The stroke impairment assessment set: Its internal consistency and predictive validity. *Archives of Physical Medicine and Rehabilitation.* 2000;81(7):863–8. DOI: 10.1053/apmr.2000.6275. 73

[233] Ang KK, Guan C, Phua KS, Wang C, Zhou L, Tang KY, et al. Brain–computer interface-based robotic end effector system for wrist and hand rehabilitation: Results of a

three-armed randomized controlled trial for chronic stroke. *Frontiers in Neuroengineering.* 2014;7(39):30. DOI: 10.3389/fneng.2014.00030. 73, 76, 77

[234] Pichiorri F, Morone G, Petti M, Toppi J, Pisotta I, Molinari M, et al. Brain–computer interface boosts motor imagery practice during stroke recovery. *Annals of Neurology.* 2015 Mar;77(5):851–65. DOI: 10.1002/ana.24390. 74, 76 , 77

[235] Kim T, Kim S, Lee B. Effects of action observational training plus BrainComputer interface-based functional electrical stimulation on paretic arm motor recovery in patient with stroke: A randomized controlled trial. *Occupational Therapy International.* 2016 Mar;23(1):39–47. DOI: 10.1002/oti.1403. 74, 76, 77

[236] Biasiucci A, Leeb R, Iturrate I, Perdikis S, Al-Khodairy A, Corbet T, et al. Brain-actuated functional electrical stimulation elicits lasting arm motor recovery after stroke. *Nature Communications.* 2018 Jun;9(1):2421. DOI: 10.1038/s41467-018-04673-z. 75, 76, 77, 104

[237] Ramos-Murguialday A, Broetz D, Rea M, Läer L, Yilmaz Ö, Brasil FL, et al. Brain-machine interface in chronic stroke rehabilitation: A controlled study. *Annals of Neurology.* 2013 Aug;74(1):100–8. DOI: 10.1002/ana.23879. 77, 78, 79, 104

[238] Ramos Murguialday A, Curado MR, Broetz D, Yilmaz Ö, Brasil FL, Liberati G, et al. Brain-machine interface in chronic stroke: Randomized trial long-term follow-up: *Neurorehabilitation and Neural Repair.* 2019 Feb;33(3):188–98. DOI: 10.1177/1545968319827573. 78

[239] Anderson KD. Targeting Recovery: Priorities of the spinal cord-injured population. *Journal of Neurotrauma.* 2004 Oct;21(10):1371–83. DOI: 10.1089/neu.2004.21.1371. 81

[240] Osuagwu BCA, Wallace L, Fraser M, Vuckovic A. Rehabilitation of hand in subacute tetraplegic patients based on brain computer interface and functional electrical stimulation: A randomised pilot study. *Journal of Neural Engineering.* 2016 Oct;13(6):065002. DOI: 10.1088/1741-2560/13/6/065002. 81, 82

[241] Donati ARC, Shokur S, Morya E, Campos DSF, Moioli RC, Gitti CM, et al. Long-term training with a brain-machine interface-based gait protocol induces partial neurological recovery in paraplegic patients. *Scientific Reports.* 2016 Aug;6(1):30383. DOI: 10.1038/srep30383. 83, 84

[242] Teng EL, McNeal DR, Kralj A, Waters RL. Electrical stimulation and feedback training: Effects on the voluntary control of paretic muscles. *Archives of Physical Medicine and Rehabilitation.* 1976 May;57(5):228–33. 85

[243] Marquez-Chin C, Bagher S, Zivanovic V, Popovic MR. Functional electrical stimulation therapy for severe hemiplegia: Randomized control trial revisited. *Canadian Journal of Occupational Therapy Revue Canadienne D'ergothérapie*. 2017 Apr;84(2):87–97. DOI: 10.1177/0008417416668370. 85

[244] Soekadar SR, Birbaumer N, Slutzky MW, Cohen LG. Brain-machine interfaces in neurorehabilitation of stroke. *Neurobiology of Disease*. 2015 Nov;83:172–9. DOI: 10.1016/j.nbd.2014.11.025. 85

[245] Popovic M, Keller T. Compex Motion: Neuroprosthesis for grasping applications. In: *Enabling Technologies: Body Image and Body Function*. Churchill Livingstone; 2003. DOI: 10.1016/B978-0-443-07247-5.50015-8. 90

[246] Marquez-Chin C, Marquis A, Popovic MR. EEG-triggered functional electrical stimulation therapy for restoring upper limb function in chronic stroke with severe hemiplegia. *Case Reports in Neurological Medicine*. 2016 Nov;2016:e9146213. DOI: 10.1155/2016/9146213. 96

[247] Jovanovic LI, Kapadia N, Lo L, Zivanovic V, Popovic MR, Marquez-Chin C. Restoration of upper-limb function after chronic severe hemiplegia: A case report on the feasibility of a brain–computer interface controlled functional electrical stimulation therapy. *American Journal of Physical Medicine & Rehabilitation* / Association of Academic Physiatrists. 2019 Feb;1. DOI: 10.1097/PHM.0000000000001163. 96

[248] Ray AM, Figueiredo TDC, López-Larraz E, Birbaumer N, Ramos-Murguialday A. Brain oscillatory activity as a biomarker of motor recovery in chronic stroke. *Human Brain Mapping*. 2020;41(5):1296–308. DOI: 10.1002/hbm.24876. 102

[249] Sebastián-Romagosa M, Udina E, Ortner R, Dinarès-Ferran J, Cho W, Murovec N, et al. EEG biomarkers related with the functional state of stroke patients. *Frontiers in Neuroscience*. 2020;14. DOI: 10.3389/fnins.2020.00582. 102

[250] Bearden TS, Cassisi JE, Pineda M. Neurofeedback training for a patient with thalamic and cortical infarctions. *Applied Psychophysiology and Biofeedback*. 2003 Sep;28(3):241–53. DOI: 10.1023/A:1024689315563. 103

[251] Kober SE, Schweiger D, Witte M, Reichert JL, Grieshofer P, Neuper C, et al. Specific effects of EEG based neurofeedback training on memory functions in post-stroke victims. *Journal of NeuroEngineering and Rehabilitation*. 2015 Dec;12(1):107. DOI: 10.1186/s12984-015-0105-6. 103, 104

[252] Nan W, Dias APB, Rosa AC. Neurofeedback training for cognitive and motor function rehabilitation in chronic stroke: Two case reports. *Frontiers in Neurology*. 2019 Jul;10. DOI: 10.3389/fneur.2019.00800. 103

[253] PhD KBC, PhD LS, PhD RRL. Neurofeedback efficacy in the treatment of a 43-year-old female stroke victim: A case study. *Journal of Neurotherapy*. 2010 May;14(2):107–21. DOI: 10.1080/10874201003772155. 103

[254] Rozelle GR, Budzynski TH. Neurotherapy for stroke rehabilitation: A single case study. *Biofeedback and Self-Regulation*. 1995 Sep;20(3):211–28. DOI: 10.1007/BF01474514. 103

[255] Ehrlich SK, Agres KR, Guan C, Cheng G. A closed-loop, music-based brain–computer interface for emotion mediation. *PloS One*. 2019;14(3):e0213516. DOI: 10.1371/journal.pone.0213516. 103

[256] Mane R, Chouhan T, Guan C. BCI for stroke rehabilitation: Motor and beyond. *Journal of Neural Engineering*. 2020 Aug;17(4):041001. DOI: 10.1088/1741-2552/aba162. 103

[257] Lim CG, Lee TS, Guan C, Fung DSS, Zhao Y, Teng SSW, et al. A brain–computer interface based attention training program for treating attention deficit hyperactivity disorder. *PloS One*. 2012 Oct;7(10):e46692. DOI: 10.1371/journal.pone.0046692. 103

[258] Qian X, Loo BRY, Castellanos FX, Liu S, Koh HL, Poh XWW, et al. Brain–computer-interface-based intervention re-normalizes brain functional network topology in children with attention deficit/hyperactivity disorder. *Translational Psychiatry*. 2018 Aug;8(1):1–11. DOI: 10.1038/s41398-018-0213-8. 103

[259] Strehl U, Aggensteiner P, Wachtlin D, Brandeis D, Albrecht B, Arana M, et al. Neurofeedback of slow cortical potentials in children with attention-deficit/hyperactivity disorder: A multicenter randomized trial controlling for unspecific effects. *Frontiers in Human Neuroscience*. 2017;11. DOI: 10.3389/fnhum.2017.00135. 103

[260] Slow cortical potentials neurofeedback in children with ADHD: Comorbidity, self-regulation and clinical outcomes 6 months after treatment in a multicenter randomized controlled trial | Cochrane Library. https://www.cochranelibrary.com/central/doi/10.1002/central/CN-01792330/full. 103

[261] Lee T-S, Goh SJA, Quek SY, Phillips R, Guan C, Cheung YB, et al. A brain–computer interface based cognitive training system for healthy elderly: A randomized control pilot study for usability and preliminary efficacy. *PloS One*. 2013 Nov;8(11):e79419. DOI: 10.1371/journal.pone.0079419. 103

[262]   Bensmaia SJ, Miller LE. Restoring sensorimotor function through intracortical interfaces: Progress and looming challenges. *Nature Reviews Neuroscience*. 2014 May;15(5):313–25. DOI: 10.1038/nrn3724. 103

[263]   Flesher SN, Collinger JL, Foldes ST, Weiss JM, Downey JE, Tyler-Kabara EC, et al. Intracortical microstimulation of human somatosensory cortex. *Science Translational Medicine*. 2016 Oct;8(361):361ra141. DOI: 10.1126/scitranslmed.aaf8083. 103

[264]   Mrachacz-Kersting N, Yao L, Gervasio S, Jiang N, Palsson TS, Nielsen TG, et al. A brain–computer-interface to combat musculoskeletal pain. In: Guger C, Allison B, Ushiba J, editors. *Brain–computer Interface Research: A State-of-the-Art Summary 5*. Cham: Springer International Publishing; 2017. pp. 123–30. DOI: 10.1007/978-3-319-57132-4_10. 103, 104

[265]   Consortium for Spinal Cord Medicine, 2005: Paralyzed Veterans of America Consortium for Spinal Cord Medicine. Preservation of upper limb function following spinal cord injury: a clinical practice guideline for health-care professionals. *Journal of Spinal Cord Medicine*. 2005;28(5):434-70. DOI: 10.1080/10790268.2005.11753844. PMID: 16869091; PMCID: PMC1808273. 22

[266]   Consortium for Spinal Cord Medicine, 1999: Consortium for Spinal Cord Medicine. Outcomes following traumatic spinal cord injury: clinical practice guidelines for health-care professionals. *Journal of Spinal Cord Medicine*. 2000 Winter;23(4):289-316. DOI: 10.1080/10790268.2000.11753539. PMID: 17536300. 22

Authors' Biographies

Cesar Marquez-Chin is a Scientist at the Kite Research Institute, Toronto Rehabilitation Institute —University Health Network. His research, informed by over two decades of working closely with clinicians, is focused on the development, testing, and improvement of practical technologies and interventions to minimize the negative impact on quality of life after paralysis. He has been part of development teams of commercial assistive devices for computer access and prosthetic systems and is a Canadian pioneer in the development of intracranial brain–computer interfaces. His work integrates brain–computer interfacing technologies into the rehabilitation of voluntary movement after spinal cord injury and stroke. He completed his doctoral studies at the Institute of Biomedical Engineering of the University of Toronto, where he is also an Assistant Professor.

Naaz Kapadia-Desai completed her doctoral studies at the Rehabilitation Sciences Institute, University of Toronto. She was the recipient of the prestigious Canadian Institutes for Health Research Doctoral Award. She is an experienced clinician and researcher with over 20 years of clinical experience and over 10 years of research experience. She currently holds a Research Associate position at KITE Research Institute, Toronto Rehabilitation Institute—University Health Network, and a Clinical Physiotherapist position with University Health Network. Her research interests include the development and validation of neurorehabilitation therapies and outcome measures specifically targeting upper extremity recovery in individuals with acquired neurological conditions. Her work explores the benefits of Functional Electrical Stimulation to restore function following SCI and Stroke. She has played a pivotal role in the commercialization of the MyndMove FES stimulator. She is an invited expert on the Spinal Cord Injury-High-Performance indicators Team that aims to develop structure, process, and outcome indicators for each prioritized domain for individuals with spinal cord injury.

Sukhvinder Kalsi-Ryan is a Clinician Scientist in the field of upper limb assessment and recovery and spine pathology at KITE Research Institute—University Health Network. Her research is oriented to establishing methods to quantify neurological change after injury and studying neuro-restorative methods to enhance and optimize function for those with neurological impairment. She is a physiotherapist by training, works as a practicing clinician, manages a research program that is focused on the intersection of discovery and clinical application and holds a doctorate degree from the Rehabilitation Sciences Institute of the University of Toronto.

Printed in the United States
by Baker & Taylor Publisher Services